Introduction to Precious Metals

Introduction to Precious Metals

Mark Grimwade

Newnes Technical Books

Newnes Technical Books

is an imprint of the Butterworth Group
which has principal offices in
London, Boston, Durban, Singapore, Sydney, Toronto, Wellington

First published 1985

© Butterworth & Co. (Publishers) Ltd, 1985

British Library Cataloguing in Publication Data

Grimwade, Mark
  Introduction to precious metals
   1. Goldwork   2. Silverwork
  I. Title
  739.2   NK7104

  ISBN 0-408-01451-2

Photoset by Butterworths Litho Preparation Department
Printed in Great Britain by Thetford Press Ltd., Thetford, Norfolk

Preface

This book introduces the fascinating subject of gold, silver and the platinum-group metals, both for those who are seeking a general insight into the subject as well as for jewellers, gold- and silversmiths and students taking courses in jewellery and decorative metalware.

After a review of the historical importance and present-day relevance and usage of the precious metals, four chapters are devoted to their occurrence, extraction and properties. Some background information on assessment of properties and their meaning is given in order that readers may appreciate the significance of the properties quoted.

Much of the remainder of the book covers the use of the precious metals for jewellery and decorative metalware. The alloys of silver, the coloured and white carat golds and the alloys of platinum are discussed and their metallurgical characteristics considered. Specific chapters deal with casting, working and annealing practice, joining processes, assaying and hallmarking, surface coatings and decoration, and electroplating. The final chapter reviews other important industrial and commercial uses for the precious metals.

A bibliography is given for those who wish to pursue any of these topics in greater detail. This does not include books dealing specifically with craft aspects or with collections in museums, etc., as these would be outside the scope of this book.

Many of the illustrations are based on ones originally drawn for a series of articles entitled 'Basic Metallurgy for Goldsmiths', published by the International Gold Corporation in *Aurum*. I am indebted to the publishers for permission to reproduce them, and to Marie-France Paroz who did the artwork.

My thanks must be recorded for the assistance given to me during the last 14 years by Brian Taylor and Chris Walton of the Worshipful Company of Goldsmiths, Pärn Taimsalu, Technical Editor of *Aurum*, and the many friends and colleagues in the Departments of Metallurgy and Jewellery and Silversmithing at the City of London Polytechnic. In particular, I acknowledge with gratitude the enormous help and encouragement given by Peter Gainsbury, Director, Design and Technology, Worshipful Company of Goldsmiths, who first introduced me to the world of the precious metals and with whom I have had many helpful discussions.

M.G.

Contents

1
Introduction

Gold, silver and platinum are all regarded as precious metals. In particular, gold and silver have been highly prized by man for thousands of years. Early civilizations associated the yellow colour of gold with the sun, and for the Egyptians, gold was the symbol of their sun god Ra. Just as gold was associated with the fire of the sun, so the brilliant white colour of silver was related to the moon. The original Latin name for silver was 'luna', meaning 'moon', although this was later changed to 'argentum', meaning white and shining.

Throughout the ages fabulous treasures have been made by artists and craftsmen exploiting the natural beauty of these metals. Alchemists spent their lives in search of the 'Philosopher's Stone' which was said to possess the property of converting base metals, i.e. the non-precious metals, to gold, and although their efforts proved to be fruitless in that direction they were instrumental in laying the foundations of the science of chemistry.

The discovery of platinum is much more recent. Although it was used by the pre-Columbian Indians of South America to make small ornaments it was not until the 18th century that its true worth was recognized. Platinum is one of a group of six metals found in association with each other and having similar characteristics. The others in the group are palladium, rhodium, iridium, ruthenium and osmium.

But what is it that makes these metals so precious? First, they have a remarkable resistance to attack by their environment. Whereas the base metals readily combine to form oxides, sulphides and other minerals, the precious metals, or 'noble metals' as they are sometimes called, can occur naturally in the uncombined state as lode deposits in rocks, or as placer deposits in the gravel beds of rivers and streams and in alluvial sands.

Secondly, gold and silver, although not platinum, were widely distributed throughout the world, which explains why so many different civilizations discovered these naturally occurring 'native metals'. It was relatively easy to extract the gold and silver, and the early metalworkers soon discerned another important property, namely, the ease with which they could be fabricated into shapes for decorative purposes thus displaying their aesthetic beauty. Nevertheless, such deposits were rare, and accordingly gold and to a lesser extent silver attained great value and became a symbol of wealth.

Unfortunately, these deposits have been largely exhausted including those found during the great gold rushes in Canada, the USA and Australia in the 19th century. Consequently, mining for the precious metals is now a large commercial undertaking.

Apart from their use for jewellery and decorative metalware, gold and silver soon assumed a monetary role because their indestructibility and rarity made them suitable for trading and for accumulating wealth. This monetary usage persisted for over 2000 years, culminating in the Gold Standard in the 19th and early 20th centuries. Whereas people in Europe, North America and Japan now tend to buy jewellery for adornment, in many other countries of the world, notably the Middle East, jewellery is purchased as an investment.

Gold has inspired artists and craftsmen throughout the ages to produce jewellery, ornaments and priceless works of art. One has only to think of the Tomb of Tutankhamun, the riches of the Ancient Greeks, Etruscans and Thracians among many others to realize the skills that these early metalworkers possessed. At a later period in the world's history, but separated from that of the Mediterranean World, the Aztecs in Mexico and the Mochica, Chimu and Inca cultures of Peru were equally adept in working gold.

Silver has also been used for artistic purposes. The Romans were excellent silversmiths as is shown by a visit to the British Museum to see the Mildenhall and Water Newton hoards, dating from the Roman occupation of Britain.

Today we find that beautiful works of art in gold, silver and platinum are being made not only using the traditional skills of the goldsmith but also incorporating modern technology in their manufacture.

In addition to their use in the decorative arts and for monetary and investment purposes, there is an ever-expanding need for the precious metals in industry and in medicine. This is discussed more fully in Chapter 15 but we can note here that since the 1950s gold has played an important role in the electronics industry; the manufacture of high-grade optical glass and control of atmospheric pollution from automobiles and industrial plants would be impossible without platinum; the photographic industry depends on silver for its existence; and the standard of dentistry has benefitted markedly from the use of all the precious metals.

The comparative rarity of the precious metals has been mentioned but what does this mean in terms of world production? Bearing in mind that there are fluctuations in the tonnages produced from year to year depending on world demand, and also that there is some disagreement in the figures quoted in various sources of reference for any one year, here are the latest figures.

Table 1.1 World output of precious metals

	Metric tonnes per annum
Gold	1300–1500
Silver	12 200
Platinum-group metals	120–130

These figures compare with an annual production of 500 million tonnes of iron and steel and 10 million tonnes of aluminium.

South Africa is the major producer of gold, accounting for about 68 per cent of total production. The USSR do not publish figures, but based on sales to non-communist countries, their output is estimated at about 20 per cent which places it second in the league, followed by Canada and the USA at about 5 per cent each.

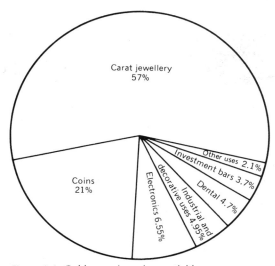

Figure 1.1. Gold usage in various activities

Figure 1.1 shows the usage of gold in 1981 in the various fields of activity. Note the staggeringly high proportion used for carat-gold jewellery. Table 1.2 shows the figures for jewellery production from various countries for the same year. This does not necessarily reflect internal consumption since some of these countries are major exporters, e.g. Italy.

Gold coins (both legal tender and bullion coins), investment bars and some jewellery are used for investment. There are two types of investors: (a) individuals who hoard gold, and (b) corporate institutions who deal in gold.

Table 1.2 Annual gold consumption in carat jewellery

	Tonnes
Europe (including Italy at 172 t)	282.4
North America	76.3
Middle East	175.4
Indian sub-continent	69.4
Far East	105.7
Latin America	27.5
Africa	16.2
Australia	3.5

For many centuries the price of gold remained virtually unchanged, and for a long period during even this century (1934–1968), the price was artificially fixed. However, by the 1960s it was evident that gold was seriously underpriced compared with other commodities, and it became necessary to free gold to find its own realistic price level. In recent years there have been wide fluctuations depending on the state of the world economy. At the time of writing (January 1984), the price is £270 a troy ounce.

The major producers of silver are Mexico (17 per cent), closely followed by the USA and Peru (16 per cent each), Canada and the USSR (11 per cent each) and Australia (7.5 per cent). A number of other countries account for the remaining 20 per cent.

Figure 1.2 shows the consumption of silver in its major categories. The value of silver is considerably less than that for the other precious metals, as

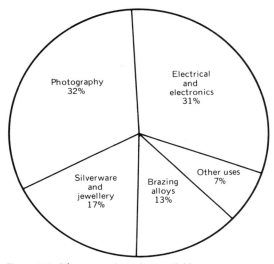

Figure 1.2. Silver usage in various activities

one might expect in view of its greater availability. It has been subject to its own price fluctuations but currently stands at £6 a troy ounce.

The total world production of the platinum-group metals includes about 87 tonnes of platinum. At present 60 per cent of the world's supply of platinum comes from South Africa, with the USSR in second place at 30 per cent and then Canada at 8 per cent.

Figure 1.3 shows the use of platinum in various fields. The high proportion used in the chemical and petroleum industries is a measure of the advantage of platinum as a catalyst. Conversely, only about 6 tonnes are used for jewellery and decorative metalware, compared with 750 tonnes of gold and about 1400 tonnes of silver every year. At present the price of platinum is similar to that of gold at £270 per troy ounce.

Throughout this book the values for weight, length, strength, etc. are generally given in SI units, i.e. tonnes, grams, millimetres, newtons per

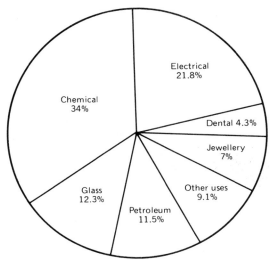

Figure 1.3. Platinum usage in various activities

square millimetre, etc. However, the weight of a precious metal is often expressed in terms of troy ounces. This system is different from the imperial system with which most of us are familiar (avoirdupois), but its use for precious metals is traditional and persists to the present. A troy ounce (oz) is equivalent to 31.1035 grams (g), and one metric tonne (t) contains 32150.72 troy oz.

2

Occurrence, extraction and refining

Gold has pride of place as the first metal to have been discovered by man. Artefacts made by hammering gold nuggets and shaping into ornaments, dating from as early as 7000 BC, have been found in Mesopotamia. Silver was probably the third metal to be found, copper having been the second. Records show that there were extensive silver workings in Cappadocia in Asia Minor during the fourth millenium BC.

In contrast, the history of the platinum metals is relatively recent. Traces of platinum must have existed in the Mediterranean area since a 7th-century-BC casket has been found in Egypt containing a piece of impure platinum, but the craftsman probably thought it was silver as the rest of the decorative work on the casket was in that metal. However, it was the Spanish conquest of the New World that revealed its use by the pre-Columbian Indians of South America for ornamental work.

The precious metals are found in the earth's crust, as are all other metallic and non-metallic elements, but in very small quantities. In 1924, Clarke and Washington published an average composition of the earth's crust based on the analysis of a vast collection of rock samples from all over the world and estimated that gold was present at between 1 and 9 milligrams (mg) per tonne, with platinum and palladium at a similar figure and silver some ten times higher. Osmium and iridium are present at about 1 mg/tonne and rhodium and ruthenium at 0.1 mg/tonne. This partly explains why these metals are so valuable: they are relatively rare compared with the base metals and the cost of extraction is very high. Even so, extraction would be extremely costly, if not uneconomic, if their distribution was uniform around the world. Fortunately, as with other important minerals, localized concentrations are found in certain regions as ore deposits and mining and extraction becomes economically feasible.

Metals are not usually found in the metallic state but rather as minerals in which they are combined with other elements such as oxygen, sulphur and silicon giving oxides, sulphides, silicates and complexes of these with other metals. Indeed, the only metals which occur in the metallic state are gold, silver, platinum and copper. The term used to describe this form of deposit is 'native' state, i.e. 'native gold', 'native silver', etc. However, even native

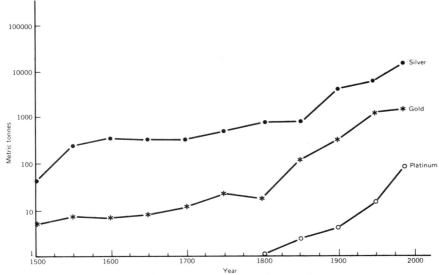

Figure 2.1. World production of the precious metals

metals are not pure since they are in combination with other metals and impurities and extensive extraction of the relevant pure metal is still required.

In spite of the known treasures of the ancient world, the production of the precious metals has risen steeply since 1500 (Figure 2.1).

Gold: occurrence and extraction

Gold is usually found as native gold in which it is combined with silver. A small amount of copper may also be present in native gold. Generally, the gold content is 85–92 per cent but can range from about 50 per cent in the very pale-yellow alloy with silver, known as 'electrum', to almost pure gold. The biggest high-quality gold nugget ever discovered was about 60 cm long and 30 cm wide and assayed at 98.66 per cent gold. The only mineral found in any quantity is gold telluride (calaverite and sylvanite).

Four types of gold deposits are found:

1. Quartz veins and lodes within rock.
2. Massive deposits in which the fine native gold particles are associated with sulphides and fine-grained quartz.
3. Disseminated copper deposits in which gold is recovered as a byproduct from copper-sulphide concentrates.
4. Placer deposits in streams or former stream gravel beds, alluvial and beach deposits.

These deposits tend to occur in folded sedimentary rocks such as are found in the Americas, Australia and South Africa.

During the great gold rushes in California, Australia and the Klondike in north-west Canada between 1848 and 1900, many thousands of gold prospectors were digging and panning for gold, but very few made their fortunes. The conditions in which they lived and worked were squalid, there was virtually no law and order, and their claims were quickly exhausted since they could only work at or near the surface. Consequently, although prospecting for gold is still done by enthusiastic amateurs as a hobby, and by the occasional professional, mining for gold now has to be a large commercial venture. A small amount of mining is done in the mountains around Dolgellau in Wales. Gold for royal wedding rings has been provided from this source.

The richest deposits are found in the Transvaal (Witwatersrand) and the Orange Free State in South Africa in a 300-mile semicircular arc. They were discovered by George Harrison in 1886. These reefs are thought to be ancient placer deposits and they exist well below the surface of the earth as conglomerates of quartz pebbles, quartz sand and native gold up to a few metres in thickness. The gold mines in this region are typically more than a mile deep, the deepest being 2.4 miles where the natural rock temperature is 50°C, and there is great danger of 'rockburst' due to the pressure exerted on the rock at these levels. Vast quantities of refrigerated air have to be pumped down to enable the workforce to mine under these conditions.

In South Africa about 75 million tonnes of ore are mined annually, producing about 30 million ounces of gold.

A number of extraction and refining techniques exist from earliest times but we shall only briefly review large-scale commercial practice. The ore is crushed and ground essentially to liberate the impure gold from the unwanted material (the 'gangue'). In some processes the crushed ore is carried down sluice boxes by water where the heavier gold particles sink down and are retained by transverse barriers called riffles due to their higher density $(16-18\,g\,cm^{-3})$ compared with that of the gangue (about $2.5\,g\,cm^{-3}$). Mercury may be added to amalgamate with the gold (and silver) and make collection easier. The mercury is removed and recovered by subsequent heating. Alternatively, the gold may be recovered by trapping it on corduroy-covered tables. In ancient times sheepskins were used for this purpose, and this was probably the origin of the legend of the Golden Fleece.

The cyanide process developed on the Witwatersrand in 1890 is used nowadays in addition to, or as a replacement for, the amalgamation stage. Finely-divided ore is treated with dilute potassium or sodium-cyanide solution into which gold, silver and copper from the ore will dissolve. After filtration to remove the gangue, zinc dust is added to the solution. The zinc replaces the gold, silver and copper in the complex cyanide and they are precipitated as metal powder. This is then melted down under oxidizing conditions to remove unwanted impurities such as copper and residual zinc as slag. For many years the final stage was the Wohlwill process in which cast bars of gold (containing silver) from the previous stage are made the anodes of an electrolytic cell containing a solution of gold chloride and hydrochloric acid. The gold is transported across the cell to be deposited as pure gold (99.95 per cent fineness or purity) on cathodes. The silver in the anode bars reacts with the solution to form silver chloride which is collected and treated separately for silver recovery. An alternative refining process, which has superseded the Wohlwill process, is the Miller process in which chlorine gas is bubbled through molten metal from the cyanide stage. Impurities, including silver, form chlorides which float to the top and are partly volatilized off or are skimmed off. The average fineness of gold produced by the Miller process is 99.7 per cent. For industrial usage, further refining to 99.9 per cent or better is desirable.

Silver: occurrence and extraction

Native silver is now much rarer than native gold as most of the silver mines of Asia Minor, Greece, Spain and the Americas have been exhausted. Only a small fraction of the world's production comes from the amalgamation and cyanidation process described in the previous section, and that largely as a byproduct of gold extraction.

Silver is mainly extracted from sulphide minerals such as argentite, tetrahedrite, polybasite, pyroargyrite and stephanite. Of these, only argentite is the simple sulphide Ag_2S, the others being complexes with antimony and copper. In addition, the silver sulphides are usually found associated with other minerals such as lead, zinc and copper sulphides so

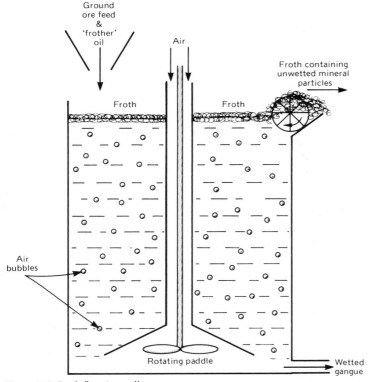

Figure 2.2. Froth flotation cell

that extraction is a complicated business and silver is a co- or byproduct in the production of these other metals.

Briefly, the crushed and finely ground ore from lode or placer deposits is treated by froth flotation to separate the minerals from the unwanted gangue (Figure 2.2). Subsequent treatment in a series of froth-flotation cells will separate the lead, zinc and copper-bearing mineral concentrates by suitable adjustment of the cell operating conditions. Each type of concentrate is then treated to remove the silver.

As an example, the lead-sulphide concentrate is smelted, i.e. refined, by roasting and reduction with coke to give lead bullion which must be desilvered. In the Parkes process, the lead bullion is remelted and 1–2 per cent zinc added. A silver–zinc alloy is formed which floats to the top and is skimmed off. This skim will contain some lead and any gold that may be present from the original ore deposit. After heating in a retort to distil off the zinc which is collected for re-use, the resultant silver-lead-gold is melted in a blast of air to oxidize the lead and other impurities forming a molten litharge (lead oxide) slag and leaving the silver (and gold) in the metallic state, known as doré metal. This process is known as 'cupellation'. The doré metal is then electrolytically refined in Moebius or Balbach-Thum cells to produce fine cathode silver (99.9 per cent purity). The anode slimes are collected for separate treatment and recovery of gold. The fine silver forms on stainless-steel cathodes as a crystalline deposit which is periodically scraped off and collected for melting to produce silver ingots.

The total world production of silver is about 400 million ounces annually.

Platinum-group metals: occurrence and extraction

The history of the discovery and development of the platinum-group metals (PGM) is fascinating, and for those who wish to know more about this I recommend the book 'A History of Platinum and its Allied Metals' by Donald McDonald and Leslie B. Hunt, published by Johnson Matthey 1982.

The Spanish conquest of South America revealed that the pre-Columbian Indians were well versed in the art of making platinum trinkets. Analysis of a 'platinum ingot' found in the Esmeraldas in Columbia gave 84.95 per cent platinum, 4.64 per cent palladium, rhodium and iridium, 6.94 per cent iron, and 1 per cent copper. Another important source was the Choco Region in Colombia. The platinum existed as small grains of white native metal together with gold grains in placer deposits in streams. It was the Spaniards who first called this white metal 'platina', a derogatory diminutive of their word for silver 'plata' and it was considered by many to be a worthless nuisance up to as recently as 1780.

Because of its high melting point it couldn't be melted and the Indians developed a technique for making ornaments which would be considered as very sophisticated even by today's standards. Small grains of platinum were coated with gold dust and heated on wood charcoal using a blowpipe. The gold melted and soldered the grains together. Further heating diffused the gold into the platinum and vice versa. The resultant coherent mass could then be hot-forged to consolidate it prior to fashioning the ornament.

The first samples arrived in Europe during the 18th century mainly through Spain, and it was here that the first attempts were made to separate-out the platinum-group metals in purer form. It was found that the impure platinum could be dissolved in aqua regia (three parts of concentrated hydrochloric acid to one part of concentrated nitric acid). Addition of sal ammoniac (ammonium chloride) precipitated out the PGM as complex chlorides leaving the iron, gold and copper in solution. After filtration, reheating the precipitate gave reversion to the metal in powder form.

About the same time French goldsmiths, notably Marc Etienne Janety (1739–1820), royal goldsmith to Louis XVI, developed the arsenic process. They showed that addition of arsenic to PGM greatly lowered the melting point so that it could be cast into clay moulds. Reheating drove off white arsenical fumes leaving a malleable platinum. Needless to say, the process was very dangerous and was not used for long.

It was the English scientists, Smithson Tennant and William Wollaston, who first isolated and reported the discovery of osmium, iridium, palladium and rhodium just after 1800. The famous French chemist Lavoisier was the first to successfully melt platinum using a stream of oxygen directed onto hollowed-out charcoal containing platinum while heating, but it was many decades later before melting could be done on a commercial scale. In the meantime, consolidation and fabrication was achieved by powder metallurgy techniques.

The next important advance was the discovery of the PGM mineral deposits at Goroblagodat and Nizhny-Tagil in the Urals, at Noril'sk-Talnakh in Siberia and later at Sudbury, Ontario in Canada. It was a Russian chemist who first reported the discovery of ruthenium in 1844. Here, PGM exists as sulphides, arsenides, antimonides and tellurides associated with copper and nickel sulphides, and hence the PGM are obtained as co- or byproducts of copper and nickel extraction.

For nearly 100 years Russia produced 93 per cent of the world's supply of PGM, but in 1924 the geologist Dr. Hans Merensky discovered the world's greatest platinum resources in an outcrop 85 miles long at Rustenburg, north-west of Johannesburg, in South Africa. This outcrop, named in his honour as the Merensky Reef, now forms part of a much larger complex (the Bushveld Igneous Complex). The platinum content is partly in the form of native metal alloyed with iron, and partly as sulphides and arsenides e.g. cooperite (PtS) and sperrylite ($PtAs_2$), together with sulphides of iron, copper and nickel. Also associated with the platinum are the other PGM. The reef is about 50 cm thick with a maximum depth of about 900 metres.

The extraction and separation of the various PGM involves a large number of complex chemical processes which are impossible to detail here. In essence, after mining, crushing, grinding, flotation and smelting, a matte containing copper, nickel and the precious metals, including some gold and silver, is produced. 'Matte' is the name given to the impure metallic sulphide product obtained from smelting sulphide ores. This matte is first treated for the extraction of copper and nickel, and the remaining sludge is further treated for the production of the precious metals. Figure 2.3 shows a simplified flowsheet for separating the PGM.

Today, the Merensky Reef supplies about 60 per cent of the world's supply of platinum, with the USSR in second place (30 per cent) and then Canada (8 per cent). Although the total PGM production of the USSR is

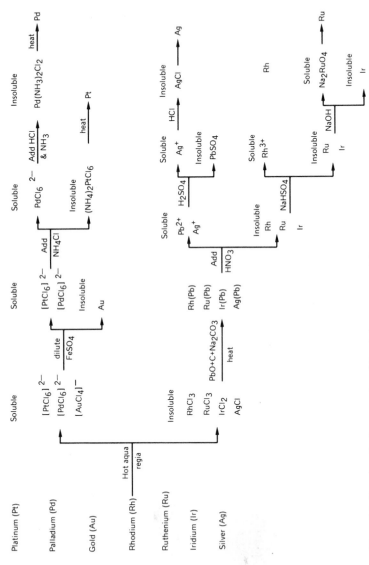

Figure 2.3. Flowsheet for the separation of the platinum-group metals

similar to that of South Africa, the platinum:palladium ratios are different, 2.5 for South Africa compared with 0.4 for the USSR; this explains the higher platinum production from South Africa. Total PGM production is about 4.5 million ounces of which 2.8 million ounces is platinum.

Recently, promising PGM deposits have been found in Montana, USA. The platinum:palladium ratio is about 0.28, and it is anticipated that in future the USA will join the major world producers, particularly of palladium.

3

Physical properties of the precious metals

The individual precious metals are elements, i.e. they cannot be split up into simpler substances. All the atoms of an element have a similar structure. There are 92 elements occurring in nature and the difference between each of them is due to their different types of atoms.

The periodic table of elements

In 1869 the Russian chemist Mendeleev published the Periodic Law which states that 'the properties of elements are in periodic dependence on their atomic weights'. Many physical and chemical properties of elements obey this law and it was possible for Mendeleev to arrange the elements in the form of a table, a modern version of which is shown in Figure 3.1. Vertical columns, known as 'Groups', contain elements showing strong family resemblances as regards their physical and chemical characteristics. For example, Group I(b) consists of the elements copper, silver and gold and although copper is regarded as a base metal, as distinct from the precious metals silver and gold, its behaviour is very similar in many respects.

Horizontal rows, known as 'Periods', also have important features in common. Note particularly that the platinum-group metals form a block in the centre of the table together with iron, cobalt and nickel and that this block is referred to as Group VIII.

Each element is numbered in ascending order when progressing through the table from top to bottom and from left to right. This is known as the 'atomic number' and it tells us something about the nature of the atoms of each element. An atom consists of a nucleus made up of electrically-positive charged particles called protons and electrically-neutral particles called neutrons together with negatively-charged particles called electrons which can be considered to orbit the nucleus in a series of layers or shells. Since the atom is electrically neutral overall, the number of protons is the same as the number of electrons and this number is the atomic number of the element. The atoms of one element therefore differ from those of another and, because many physical and chemical properties depend on the atomic structure, these characteristics vary from one element to another. The periodicity discovered by Mendeleev and others arises from the fact that

14

Group	I a	II a	III a	IV a	V a	VII a		VIII		I b	II b	III b	IV b	V b	VI b	VII b	0
Period 1																1 H	2 He
2	3 Li	4 Be										5 B	6 C	7 N	8 O	9 F	10 Ne
3	11 Na	12 Mg										13 Al	14 Si	15 P	16 S	17 Cl	18 A
4	19 K	20 Ca	21 Sc	22 Ti	23 V	24 Cr	25 Mn	26 Fe	27 Ni	28 Co	29 Cu	30 Zn	31 Ga	32 Ge	33 As	34 Se	35 Br	36 Kr
5	37 Rb	38 Sr	39 Y	40 Zr	41 Nb	42 Mo	43 Tc	44 Ru	45 Rh	46 Pd	47 Ag	48 Cd	49 In	50 Sn	51 Sb	52 Te	53 I	54 Xe
6	55 Cs	56 Ba	57-71 *R.E	72 Hf	73 Ta	74 W	75 Re	76 Os	77 Ir	78 Pt	79 Au	80 Hg	81 Tl	82 Pb	83 Bi	84 Po	85 At	86 Rn
7	87 Fr	88 Ra	89 Ac	90 Th	91 Pa	92 U	Transuranic elements											

* Rare earth elements

Figure 3.1. The Periodic Table of Elements

similarities in electron orbit configuration periodically re-occur with ascending atomic number, hence the similarities in certain properties in elements within a group.

It will be apparent that there should be an increase in atomic weight with atomic number since the numbers of protons, electrons and incidentally also neutrons are increasing, and almost without exception this is found to be true. The atomic weight is on a comparative scale using the standard of

Table 3.1 Atomic characteristics of the precious metals

Element	Chemical symbol	Atomic number	Atomic weight	Atomic radius (nm*)
Silver	Ag	47	107.87	0.1443
Gold	Au	79	196.97	0.1442
Ruthenium	Ru	44	101.07	0.1340
Rhodium	Rh	45	102.91	0.1369
Palladium	Pd	46	106.4	0.1347
Osmium	Os	76	190.2	0.1337
Iridium	Ir	77	192.2	0.1356
Platinum	Pt	78	195.09	0.1386

* One nanometre (nm) = 10^{-9} metres

16 as the weight of an oxygen atom. Table 3.1 gives the atomic numbers and weights of the precious metals together with their atomic radii. The alloying behaviour of metallic elements is partly determined by the atomic radii and the atomic size mismatch of the elements present in the alloy. These factors also influence the strength of the alloy.

Crystal structure

Atoms may be regarded for simplicity as very small spheres, the actual diameters of which depend on the individual element. Obviously, to make a lump of metal which can be handled, a very large number of atoms must be brought together and bonded in some way.

In a quantity of liquid, the atoms are loosely held together in a random fashion and move about relatively easily. When the liquid freezes to form a solid, all the atoms take up positions in a regular geometric array known as a lattice to form the crystal structure characteristic of that particular substance. A two-dimensional representation is shown in Figure 3.2.

All true solids have a regular crystal lattice whether they are metals or non-metals. The main factor which distinguishes metals from non-metals is

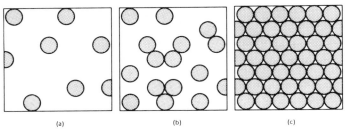

Figure 3.2. Atomic arrangement in (a) gases, (b) liquids, and (c) solids

the manner in which their atoms are bonded in the lattice. Space does not permit a review of all the bonding mechanisms in solids, but metals have a special form of bonding (metallic bonding) in which the atoms are not so tightly held as for non-metallic solids and in which some of their outermost electrons 'float' throughout the lattice as an 'electron cloud'. It is this type of bonding which is responsible for the characteristics of ductility, toughness and electrical conductivity which one associates with metals.

There are 14 ways in which spheres can be stacked to form a regular geometric array. However, the vast majority of metals fall into one of three categories. Furthermore, the precious metals, with the exceptions of

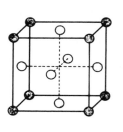
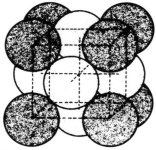

Figure 3.3. The face-centred cubic cell

osmium and ruthenium, have a lattice in which the unit cell or basic building block is face-centred cubic (FCC), i.e. an atom at each corner of the cube and one in the centre of each face of the cube (Figure 3.3). If other similar cells are added on in all three directions ad infinitum, a metal crystal can be built up which is sufficiently large enough to handle (Figure 3.4). It will be noticed that the corner and face atoms are shared between adjacent cells.

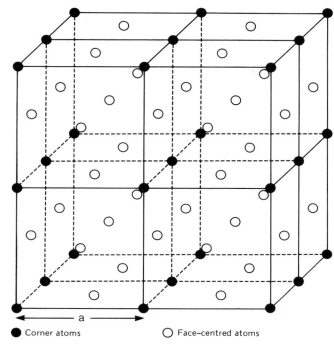

● Corner atoms   ○ Face–centred atoms

Figure 3.4. Block of eight face-centred cubic cells

Although the face-centred cubic lattice is by no means confined to the precious metals – copper, nickel, lead and aluminium among others also have this structure – it is noteworthy that the high ductility and malleability for which the precious metals are well known largely results from this particular type of lattice. Osmium and ruthenium are exceptions in that their lattice is close-packed hexagonal (Figure 3.5) which often displays limited ductility. Indeed, osmium and ruthenium are notoriously difficult to fabricate, and consequently their industrial uses are somewhat limited.

Although several metals have the same crystal structure, the dimensions of the unit cell, i.e. the lattice parameter, is different in each case since the atomic radii vary (Table 3.2).

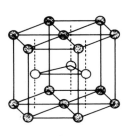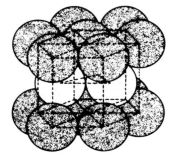

Figure 3.5. The close-packed hexagonal cell

Density

The density of a substance is defined as its weight per unit volume and it is usually expressed in $g\,cm^{-3}$, although the preferred SI unit is $kg\,m^{-3}$. Density is a function of both atomic weight and crystal structure. Therefore, elements with high atomic weights usually have high density particularly if their atoms are closely packed together in the lattice. Both the FCC and HCP structures represent the closest form of atomic packing of all types of lattice.

Table 3.2 Crystal structure and lattice parameters

Metal	Structure	Lattice parameter (nm)	
Gold	FCC	0.40786	
Silver	FCC	0.40862	
Platinum	FCC	0.39229	
Rhodium	FCC	0.38029	
Palladium	FCC	0.38906	
Iridium	FCC	0.38392	
		a	c*
Osmium	HCP	0.27340	0.43194
Ruthenium	HCP	0.27056	0.42816

* The close-packed hexagonal (HCP) cell has two characteristic dimensions and hence two parameters must be stated. The parameter for the FCC cell is the length of the cube edge.

Table 3.3 Density values for the precious metals at 20°C

Metal	Density $(g\,cm^{-3})$
Gold	19.32
Silver	10.49
Platinum	21.45
Palladium	12.02
Rhodium	12.41
Iridium	22.65
Osmium	22.61
Ruthenium	12.45

Consequently it is not surprising to find that gold, osmium, iridium and platinum are regarded as heavy metals (Table 3.3). Temperature also affects the density value because the crystal structure expands on heating. Density values are usually quoted for measurements made at 0°C or 20°C on fully-annealed metal.

Melting and boiling points

The melting temperatures of gold, silver, platinum and palladium are of special interest and of great practical importance because they are used as primary and secondary fixed points on the International Practical

Table 3.4 Melting and boiling points of the precious metals

Metal	Melting point (°C)	Boiling point (°C)
Gold	1064.59	2808
Silver	960.5	2187
Platinum	1769	3827
Palladium	1552	3167
Rhodium	1960	3877
Iridium	2443	4800
Osmium	3050	5300
Ruthenium	2310	4150

Temperature Scale from which all temperature-measuring devices are callibrated. This scale is periodically revised, the latest being 1975, as standards of purity and measuring techniques are improved. The precious metals are chosen as reference points because they are produced to a high degree of purity and are free from contamination by oxidation.

Boiling points are more difficult to determine and a wide variation in values has been reported in the technical literature over the years. The values quoted in Table 3.4 are the currently-accepted values.

Optical properties

Much of the fascination of gold throughout the ages has been due to its beautiful yellow colour coupled with its high lustre or reflectivity which is retained indefinitely because of its resistance to corrosive attack by the environment. Copper is the only other metal with a characteristic colour, but unfortunately it lacks the corrosion resistance of gold. Similarly, the aesthetic appeal of silver largely arises from its brilliant white colour and very high reflectivity. As we shall see in Chapter 6, the tarnishing behaviour of silver limits its usefulness as a reflector.

The colours of gold and copper result from sharp changes in reflectivity with the wavelength of light λ, reflectivity being low at the blue end of the visible light spectrum, i.e. λ less than 500 nm but rising dramatically at the yellow—red—infra-red end of the spectrum (500–1200 nm). This change in reflectivity is associated with the nature of the electron configurations in the atomic structures of gold and copper, but a detailed discussion is beyond the scope of this book.

Figure 3.6 shows the relationship between per cent reflectivity and wavelength of light for gold, silver and some of the platinum group metals. Notice that silver has a high reflectivity over the entire visible light spectrum which is responsible for its brilliant white lustre. Of the platinum metals, rhodium has the highest reflectivity with a mean value of about 80 per cent which, coupled with a lack of tarnishing behaviour, makes it eminently suitable as an electroplated finish for silver.

The optical properties of gold are profoundly altered by alloying, and this is of considerable importance to the jeweller and goldsmith. The effect of alloying on the colour of carat golds will be discussed more fully in Chapter 7.

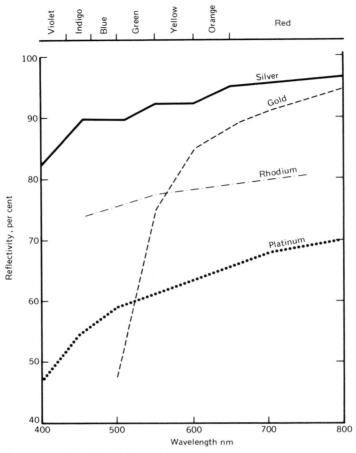

Figure 3.6. Reflectivity of the precious metals

It is possible to make gold films and coatings which are thin enough to be translucent. Gold leaf about 100 nm thick and thin films of gold, 5–40 nm thick, on glass will transmit substantial amounts of green light. Gold deposits of thickness less than 4 nm transmit light which may be grey, yellow, red or violet depending on the actual thickness and method of deposition. Deposited gold films will reflect infra-red radiation. This combination of transmission and reflectance properties makes thin gold deposits suitable for glass windows of offices and automobiles, optical filters, solar-energy collectors and heat reflectors.

Other physical properties

So far we have considered the important physical properties that influence alloying behaviour, fabrication and uses of the precious metals. Table 3.5 draws attention to other physical characteristics. Particularly noteworthy is

Table 3.5 Other physical properties of the precious metals

Metal	Resistivity at 0°C	Temperature coefficient of resistance over 0–100°C	Thermal conductivity 0–100°C	Coefficient of expansion over 0–100°C	Mean specific heat over 0–100°C
	$\mu\Omega$ cm	$10^{-3}/°C$	W/m K	$10^{-6}/K$	J/kg K
Au	2.01	4.0	315.5	14.1	130
Ag	1.47	4.1	425.0	19.1	234
Pt	9.85	3.92	73	9.0	134.4
Pd	9.93	4.2	76	11.0	247
Rh	4.33	4.4	150	8.5	243
Ir	4.71	4.5	148	6.8	130.6
Os	8.12	4.1	87	6.1	130
Ru	6.80	4.1	105	9.6	234

the fact that of all metals silver has the highest electrical and thermal conductivity. The relatively high electrical conductivity of gold coupled with its excellent tarnish and corrosion resistance has meant that it is widely used in low-voltage solid-state devices as an electrical contact material where high contact loadings cannot be used to break tarnish films.

4

Mechanical properties of the precious metals

The mechanical properties of metals and alloys are of particular interest to engineers, designers, manufacturers and, in the case of the precious metals, jewellers and silversmiths because they give valuable information on how an article will behave during fabrication, and hence in the design of processing schedules, and also on how the article or component will stand up to the service stresses imposed on it during its life. The relevant properties are hardness, yield and tensile strength, ductility and malleability, toughness and sheet-metal formability. These properties can be profoundly affected by alloying, presence of impurities, temperature, condition of the metal, i.e. whether it is annealed or has been hardened by a prior working operation, special heat treatments which influence the internal microstructure particularly of certain alloys such as some carat golds, the stress systems which may be operating during fabrication and in service and even the presence of a corrosive environment.

The mechanical properties of pure gold, silver and the platinum-group metals will be listed here but the influences of alloying, heat treatments etc. will be given in the relevant subsequent chapters.

Hardness

The hardness of a material is measured by determining its resistance to penetration by an indenter of standard specified shape and dimensions applied using specified conditions of loading. Hardness is a useful property in that it gives some indication of the resistance of a material to damage by wear and scratching. It is related also to the strength of the material and may give an approximate indication of its working behaviour. High hardness values are associated with greater strength, and in many cases to improved wear resistance.

It is sometimes stated that high hardness is related to a reduction in ductility. While this is true for any particular metal or alloy which has been cold-worked or perhaps heat-treated to increase its strength, it is not necessarily true when different materials are being compared in, say, the annealed condition.

It is a relatively easy test to perform using a standard hardness-testing machine. Several different methods are available, the three most commonly used being the Vickers, the Brinell and the Rockwell tests. Because the test conditions are not the same for each method, it is unwise to compare or to convert hardness values from one scale to another, although for relatively soft materials the Vickers hardness numbers (HV) and the Brinell numbers (HB) are somewhat similar.

Space does not permit a description of all three techniques, and so only a brief outline will be given of the Vickers test which has the important advantage over the other methods in that the indentations are smaller and

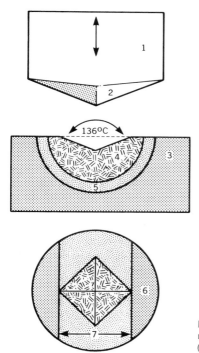

Sketch of indentation viewed
through the microscope

Figure 4.1. The Vickers Hardness Test. (1) Steel mount. (2) Square-based diamond pyramid indenter. (3) Test specimen. (4) Region of plastic deformation. (5) Region of elastic deformation. (6) Shutter. (7) Measure length of diagonals

the same indenter is used for measuring a wide range of hardness values from those of soft materials such as gold to those of very hard materials. It is also the most suitable method for the precious metals.

The indenter is a square-based pyramidal diamond of specified shape and dimensions, the apex of which is forced into the test-piece surface using a standard load P measured in kilograms of force (kgf) and which can vary from 1 kgf to 100 kgf depending on the size of indentation required and the hardness of the specimen being tested. After a set time of 15 seconds, the indenter is removed. The two diagonals of the square indentation are measured in millimetres using a microscope with an eyepiece containing a shutter system and a graduated scale (Figure 4.1). The measurements from a minimum of at least three indentations are then averaged, the average surface area A of the impressions is calculated and the Vickers Hardness Number HV is given by

$$HV = \frac{P}{A} \text{ kgf mm}^{-2}$$

In practice, tables of values are provided and this removes the need for calculating A. The HV value can be looked up directly from the tables once the average length of the diagonal has been measured.

The modern convention is to use the symbols HV, HB (Brinell Hardness) and HRA, HRB, HRC (depending on the Rockwell scale used) for the three systems, although in older books readers may find VPN, DPN, VPH or DPH given for the Vickers test and BHN for the Brinell test.

Strength

Strength is generally regarded as the capacity of a material to resist either a permanent distortion or fracture when a force is applied to it. Strength is assessed in a tensile test using a machine in which the extension of a test specimen is measured as a steadily increasing tensile load is applied (Figure 4.2).

When a test specimen of the material is stretched under tension with an increasing load or force P, a point is eventually reached where the material will start to deform plastically. This is known as 'yielding'. At load values below the yield load P_L, the test specimen deforms elastically, and on removal of the load it will revert to its original dimensions. Above the yield

Figure 4.2. A tensile-testing machine. Precious metals are usually tested in wire or strip form using a smaller machine to reduce material costs

load, however, there will be a permanent change in dimensions due to plastic deformation. As the load is increased further, the test specimen continues to extend in the direction in which the load is applied and the cross-sectional area decreases uniformly along its length until the point is reached where the area continued to decrease only in a localized region. This phenomenon is known as 'necking' (Figure 4.3) and it begins when the load has reached its maximum value P_{max}. Upon stretching the test bar still further, the applied load decreases as necking continues until fracture occurs.

Figure 4.3. 'Necking' in a tensile-test specimen

This behaviour may be shown graphically by a load-extension curve (Figure 4.4). The recommended units of load or force are newtons (N) or multiples such as kilonewtons (1 kN = 1000 N), meganewtons (1 MN = 10^6 N), etc.

It is obvious that the values of P_L and P_{max} will be greater for a thick bar specimen than for a thin wire. It is preferable, therefore, to divide the value of load by the original cross-sectional area A_0 of the test specimen in order to obtain a value for the yield strength and tensile strength of a material which is independent of the dimensions of the specimen. This means that we can express strength as a 'stress':

$$\text{Stress} = \frac{\text{Load}}{\text{Area}} = \frac{P}{A_0} \ \text{N mm}^{-2}$$

The units of megapascals are often used where $1 \text{ MPa} = 1 \text{ N mm}^{-2}$.

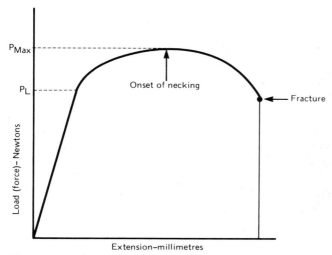

Figure 4.4. Load versus extension curve

Similarly, the extension of the specimen during application of the load may be expressed as a 'strain' by dividing the extension $(L_1 - L_0)$ by the original length L_0:

$$\text{Strain} = \frac{L_1 - L_0}{L_0}$$

You will notice that strain has no units (it is a dimensionless quantity) because the units of length cancel out from the top and the bottom of the equation.

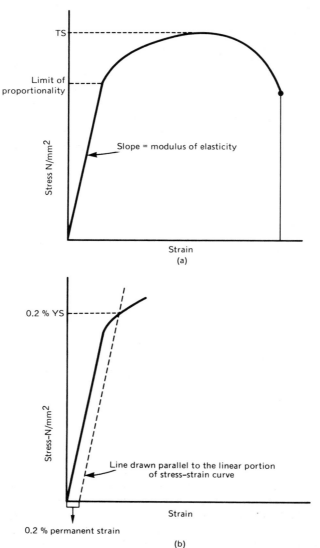

Figure 4.5. (a) Stress versus strain curve. (b) Determination of 0.2 per cent proof stress

The load-extension curve may be replotted as a stress-strain curve for the material (Figure 4.5a), and from it the following tensile properties can be determined.

Yield strength (YS)

In many respects this is a very important property because it is the stress at which a material starts to deform plastically. It must be exceeded during any forming process in order to achieve a permanent change of shape, e.g. production of strip, rod, wire tube, etc. However, once an article or component or structure has been fabricated, further deformation is undesirable and any stress met when the article is in service must not exceed the yield strength of the material.

In practice, the exact value of yield stress is difficult to measure because strictly speaking the load P_L measures the proportional limit which is slightly below the actual yield load. Consequently, the limit of proportionality $= P_L/A_0\,Nmm^{-2}$ and, say, 0.2 per cent proof stress (PS) are often used to indicate the yield strength of a material. To determine proof stress, the value of stress to produce a very small amount (e.g. 0.2 per cent) of permanent strain is measured accurately. This value of stress is known as the 0.2 per cent PS (Figure 4.5b).

Tensile strength (TS)

This is a convenient property to report as it indicates the maximum strength of a material, i.e. it is a measure of the maximum load which can be applied without the occurrence of necking and fracture.

$$TS = \frac{P_{max}}{A_0}\ N\,mm^{-2}$$

Modulus of elasticity (E)

We have seen from Figures 4.4 and 4.5 that the plot is linear up to the yield load or proportional limit P_L, i.e. 'stress is proportional to strain' which is a statement of the law propounded by Robert Hooke (1635–1703) for the elastic behaviour of a material. The physicist Thomas Young restated this mathematically as

$$\frac{Stress}{Strain} = a\ constant = E$$

The value of the constant E is known as the Modulus of Elasticity of the material and it is measured by calculating the slope of the straightline portion of Figure 4.5. It does not indicate the strength of a material but, nevertheless, it is an important engineering property because it is a measure of stiffness. The values for randomly oriented polycrystalline specimens, which is a normal condition for metallic materials, vary for different pure metals but are not usually greatly affected by alloying.

Ductility

The ductility of a material is an indication of its ability to deform plastically, and therefore a high value of ductility is associated with good formability or malleability. Two assessments of ductility can be obtained from a tensile test:

(a) Elongation to failure, sometimes referred to simply as elongation and expressed as a percentage:

$$\text{Elongation per cent} = \frac{L_f - L_0}{L_0} \times 100$$

where L_0 is the original length (gauge length) and L_f is the length at failure. The value of L_0 should always be stated because as a result of localized necking, the elongation value varies with gauge length for a given cross-section area of the specimen.

(b) Reduction in area, defined as

$$\text{RA per cent} = \frac{A_0 - A_f}{A_0} \times 100$$

For theoretical reasons, the RA value is a better measure of the formability or malleability of a metal, e.g. the RA value of pure gold is almost 100 per cent whereas a typical elongation value is only 50 per cent. Annealed mild steel has a similar elongation value to that of gold but its RA value is also about 50 per cent and it certainly cannot be beaten into thin sheet and leaf as gold can.

Toughness

It is a mistake to assume that good toughness means the same as high strength or high ductility. Toughness may be defined as the ability of a material to absorb energy during plastic deformation up to the point of

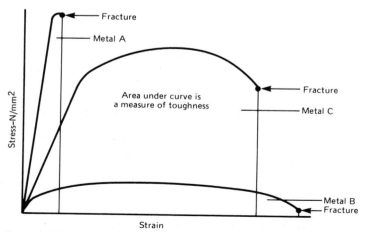

Figure 4.6. Stress-strain curves for three hypothetical metals

fracture. For example, Figure 4.6 shows the stress-strain curves for three hypothetical metals. Metal A has a high tensile strength but fractures with virtually no permanent strain or deformation. In other words, it is strong but brittle. Metal B has a high ductility but a relatively low tensile strength and it cannot be considered as displaying good toughness. The pure precious metals can be said to fall into the category of Metal B.

The area under the stress-strain curve is a good indication of toughness and it can be seen that toughness is linked to a combination of strength and ductility such as is shown by Metal C. Some of the precious metal alloys show this behaviour, a good example being the nickel white carat golds discussed in Chapter 4. However, it is not common practice to assess toughness in routine tensile tests.

Sheet-metal formability

The tensile and hardness tests and measurements of ductility do not give a full picture of how a metal is likely to deform during certain working processes. This is particularly true for sheet-metal-forming operations, and since precious metals in sheet form are often used for fabricating articles it is common practice to assess the cupping or deep drawing characteristics of the metal. One test which is widely used is the Erichsen cupping test.

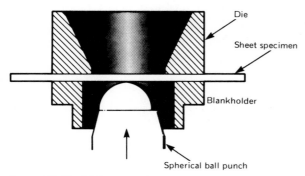

Figure 4.7. The Erichsen cupping test

In this test an annealed sample of the sheet is clamped between a die and a blankholder using a standard load. A spherical punch of 20 mm diameter is pushed into the sheet so that it stretches over the punch and forms a cup-like depression (Figures 4.7 and 4.8). Eventually the sample fractures, and the depth of the cup at this point is recorded as the Erichsen value in millimetres.

Care should be taken in interpreting the data since the test only relates to stretching over a former. Nevertheless, the test may be useful in making a comparison between different alloys.

Other properties and tests

A metallic material possesses a number of properties which can be assessed by suitable tests.

Elevated temperature tensile properties

The various tensile properties described in this chapter can be assessed over a range of temperatures simply by placing a furnace around the test specimen in the testing machine. Such tests would be performed if the material is in service at temperatures other than the ambient temperature.

Impact strength and fracture toughness

The precious metals and their alloys generally display good impact resistance and fracture toughness, i.e. they do not fracture in a brittle

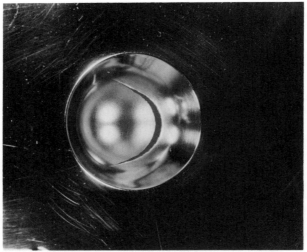

Figure 4.8. Erichsen cup failure in sterling silver

manner when a force is applied rapidly. However, the presence of small amounts of harmful impurities can have a disastrous effect. For example, as little as 0.2 per cent lead in the carat-gold alloys will render them brittle.

Creep

Metals will slowly plastically deform over very long periods of time at high temperatures if a stress is applied, even if this stress is below the yield stress. This is known as 'creep' and it may be necessary to assess creep resistance if a metal or alloy is continually in service at elevated temperatures.

Fatigue

Metal fatigue is a common cause of failure in manufactured articles and components which are subjected to cyclic loading conditions in service. Examples are wires in the flexible bracelets of wrist watches and brooch pins and clips. Fracture occurs as a result of the repetition of a sufficiently large number of applications of stress which is insufficient to cause failure upon a single application.

Fatigue tests are costly to perform with the precious metals because of the type of specimen required, and are rarely done.

Cold-working

Cold-working a metal or alloy greatly affects its mechanical properties. Hardness, yield and tensile strength increase with increasing amounts of cold work, i.e. deformation at room and relatively low temperatures, and there is a corresponding decrease in ductility (Figure 4.9). This is known as

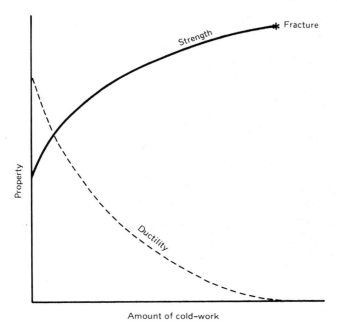

Figure 4.9. Effect of cold-work on properties

'work hardening' and it is easily demonstrated by taking a thin piece of metal such as an opened paper clip and bending it backwards and forwards. It becomes increasingly harder to bend the metal and eventually fracture occurs.

The ductility may be restored and the cold-worked strength reduced to the original level by heating the metal at a sufficiently high temperature to recrystallize it – a process known as annealing. This is discussed more fully in Chapter 10.

The precious metals

Table 4.1 lists the mechanical properties of gold, silver and platinum-group metals. Values are given for metals in both the annealed and cold-worked

Table 4.1

Metal	Condition	Hardness HV	Yield strength N mm^{-2}	Tensile strength N mm^{-2}	Modulus of elasticity N mm^{-2}	Elongation of 50 mm G.L. %	Reduction in area %	Erichsen Cup depth mm
Gold	Annealed	20	23	124	7.72×10^4	45	100	
	Hard	58	207	216		4		
Silver	Annealed	22		172	8.06×10^4	50	'	8.4
	Hard	100		375		5		
Platinum	Annealed	37		124	17.2×10^4	45		12.2
	Hard	108		349		7.5		
Palladium	Annealed	37		172	11.7×10^4	40		
	Hard	106						
Rhodium	Annealed	120		689	31.7×10^4			
	Hard	260						
Iridium	Annealed	220	2000	1103	51.6×10^4			
	Hard							
Ruthenium	Annealed	220		496	41.4×10^4	10		
	Hard	360/750						
Osmium	Annealed	400			55.8×10^4			

conditions. For the annealed metals, slight variations are found in the technical literature because of variations in the standard of purity, annealing temperature, etc., but the values quoted are generally accepted. The values for the cold-worked condition depend not only on purity but also on the amount of cold work received as shown in the previous section. It is common practice to quote values for about 60 per cent cold work, i.e. a 60 per cent reduction in thickness in sheet by rolling or a 60 per cent reduction in area by wire drawing or rod rolling.

5

Chemical properties of the precious metals

Precious metals have excellent resistance to chemical attack (i.e. corrosion resistance) although, as we shall see, this is only true up to a point. Most metals react in some way or other with their surroundings. Some react with oxygen in air to form a stable oxide on their surface – a phenomenon enhanced by heat. Again, most metals dissolve in acids to form metallic salts or corrode in solutions of salts. Many are affected by contact with mercury with which they form substances called amalgams.

Chemical attack or corrosion is basically an electrochemical process. The metal being attacked behaves in a similar manner to an electrode in an electrolytic cell or battery. When a metal comes into contact with certain

Table 5.1 The electrochemical series

Metal	$E°$ volts	Metal	$E°$ volts
Gold	+1.68	Ruthenium	+0.45
Platinum	+1.20	Copper	+0.34
Iridium	+1.0	Nickel	−0.23
Palladium	+0.83	Iron	−0.41
Rhodium	+0.80	Zinc	−0.76
Silver	+0.80	Aluminium	−1.70
Osmium	+0.7	Magnesium	−2.37

substances, whether potentially corrosive or not, a voltage difference is created between the two. The value of the voltage depends on the particular metal-chemical system involved, but by comparing it with that for a standard electrode, in practice the standard hydrogen electrode, each metal can be assigned a unique value known as the standard electrode potential ($E°$). The value for some metals are shown in the Electrochemical Series (Table 5.1).

The more positive the value, the more resistant is the metal to attack: gold is at the top of this list followed by the other precious metals. It is because of their position in the series that these metals are called the 'Noble Metals' and are noted for their corrosion resistance.

Chemical behaviour of gold

The atomic number of gold is 79, and 20 isotopes are known. Isotopes are forms of a given element in which the atoms have the same numbers of protons and electrons but the numbers of neutrons will vary between each isotope. The chemical properties are identical but the atomic weights are different. If the isotope is unstable, its atomic nucleus will revert to a stable form and emit radiation in doing so, i.e it is a radioactive isotope. Only one stable non-radioactive isotope, Au^{197}, exists and this, of course, is the form of gold with which we are familiar. In terms of its chemical compounds gold exhibits either the monovalent or trivalent state. Valency is measured by the number of atomic weights of hydrogen with which one atomic weight of the element, in this case gold, will combine, hence monovalent, divalent, etc.

Gold is totally unaffected by exposure to wet and dry atmospheres and it does not react with air or oxygen even at high temperatures. It is for this reason that gold is usually found in its free metallic state, 'native gold'.

It is insoluble in sulphuric acid (H_2SO_4), nitric acid (HNO_3) and hydrochloric acid (HCl). However, it will dissolve in aqua regia which is a mixture of three parts of concentrated HCl to one part of concentrated HNO_3 giving a solution of gold chloride $H(AuCl_4)$. This in itself is highly corrosive, and dissolution has to be done in vessels lined with either glass or the metal tantalum. It is said that the name aqua regia, meaning 'royal water', arose because it is the only acid which will dissolve gold.

Gold will dissolve in cyanide solutions. This property is exploited by the gold-mining industry for extracting gold from crushed and finely ground ore.

Under certain special conditions gold may dissolve in solutions of chlorides and other halides. If hydrogen-sulphide gas is bubbled through gold-chloride solution, gold sulphide is produced.

When put into contact, gold and mercury combine to form an amalgam that may be pasty or liquid depending on the relative proportions of the two metals. Since mercury evaporates at a relatively low temperature, it can be driven off by gentle heating leaving the gold. This must be done in well-ventilated conditions as mercury vapour is highly toxic. Jewellers are sometimes asked by customers why gold ring shanks have turned white and it is usually found that the person has handled a broken mercury thermometer in the home. Gentle heating with a low flame will restore the gold colour without damage to any mounted gemstone provided the temperature is not allowed to rise more than about 150°C.

If a suitable reducing agent, e.g. hydrazine hydrate, is added to gold-chloride solution, an extremely finely-divided sub-microscopic form of gold is produced known as colloidal gold, which gives a red, blue or violet colouration to the solution. Similarly, if a solution of stannous chloride is added to gold chloride, a precipitate of stannic oxide coloured purple by the presence of colloidal gold is obtained. This is known as 'Purple of Cassius' and it is used as a colouring agent for enamels and glasses. One part of gold in 50 000 parts of glass gives a ruby-red colour. This is also a very sensitive rapid colour test for the presence of small amounts of gold in substances.

Chemical behaviour of silver

The atomic number of silver is 47, and of the 27 isotopes two are the stable non-radioactive forms found in nature, Ag^{107} (51.82 per cent) and Ag^{109} (48.18 per cent). Silver is generally monovalent although other valencies are known to exist.

Silver does not react to any great extent with wet or dry air as such but it can form unstable oxides under certain conditions which revert to silver on heating with the loss of oxygen. The presence of hydrogen sulphide in moist air leads to the formation of silver-sulphide films on the surface of silver. This is the cause of tarnishing which is discussed more fully in the next chapter.

One remarkable property of silver is the fact that in the molten state it can dissolve vast quantities of oxygen, e.g. 100 g of silver will dissolve 213.5 ml of oxygen equivalent to 0.305 per cent by weight at 973°C. During solidification, the dissolved oxygen comes out of solution and reforms as molecular gas which either is trapped as gas blowholes, or causes a pronounced spitting action. Bullion dealers have regarded the degree of 'silver spit' as a rough measure of the purity of silver. If copper or any other base metals are present they will combine with the oxygen to form solid oxides and the amount of spit is reduced.

Unlike gold, silver will dissolve in nitric acid and in hot concentrated sulphuric and hydrochloric acids. This fact is made use of in gold assaying in which silver is 'parted' from gold by solution in HNO_3.

The silver halides, e.g. silver chloride AgCl and silver bromide AgBr, decompose to metallic silver when exposed to visible or ultra-violet light and this is the basis of the photographic industry.

Silver possesses one other remarkable property for which the name 'oligodynamic effect' has been coined from the Greek words 'oligos' meaning small, and 'dynamis' meaning power, i.e. effective in small amounts. This is concerned with the killing effect in bacteria. It has long been known that water contained in silver vessels remains purer for longer periods than when stored in earthenware or other containers. It has been recorded that Alexander the Great would only drink water out of his silver helmet. The sterilizing power of silver has been proved scientifically and today it is used for silver skull plates, as colloidal silver for treatment of wounds and for the disinfection of water, ice, fruit juices, etc. This behaviour is of special interest to the silverware trade as the use of solid silver or electroplated tableware is desirable from a hygienic standpoint.

Chemical behaviour of the platinum-group metals

The platinum-group metals, with the exception of platinum itself, will react with air and oxygen on heating forming surface oxides which discolour the metals. However, upon heating above a certain critical temperature, depending on the particular metal, the oxides volatilize off leaving clean surfaces.

They are insoluble in most acids but platinum and palladium dissolve in aqua regia. The other platinum-group metals are largely insoluble in aqua regia. This difference is utilized in the early stages of the separation of the six metals. Palladium will also dissolve in boiling concentrated sulphuric acid.

Palladium will absorb vast quantities of hydrogen gas and this property is of great industrial importance in the use of palladium diffusion units for the production of pure hydrogen.

Although strictly speaking this is not a chemical property, platinum has a very valuable function in chemical engineering. Many chemical reactions can proceed at a very rapid rate in the presence of a platinum surface whereas they would be uneconomically slow in its absence. This behaviour is known as catalysis and platinum stands supreme among all metals as a catalyst. The catalyst remains unchanged throughout the reaction.

Finally, it is worthy of mention that chemical compounds of platinum, the drugs cisplatin and neoplatin, are playing an important role in chemotherapy treatment of cancer.

6
Silver and its alloys

To review the properties and uses of silver and all its important alloys is an enormous task and impossible for a book of this nature. The same can be said for the other precious metals and their alloys. For those readers who are interested in studying all aspects of the subject, a bibliography is given at the back of this book. Consequently, this and all of the following chapters, apart from Chapter 15, are mainly concerned with the use of these metals and their alloys by goldsmiths and silversmiths in the manufacture of jewellery, goldware and silverware.

Silver

Silver in its commercially pure form is often referred to as 'fine silver' and by definition it contains at least 99.9 per cent silver (Ag), the principal impurities being copper (Cu), lead (Pb) and iron (Fe). 'High fine' silver contains at least 99.95 per cent Ag and 'spectroscopically pure' silver is 99.999 per cent Ag. However, for the purposes of alloy manufacture, commercial fine silver is used. Fineness is used to describe how much precious metal is present in the metal or alloy in terms of so many parts in 1000, i.e. commercially pure silver is of a fineness not below 999.

Although fine silver was used by many civilizations from earliest times, principally for monetary purposes, it was soon recognized that it lacked the necessary strength for use in making artefacts. Wear resistance is poor and such silverware is easily dented and destroyed. Generally, additions of other metals to produce an alloy considerably increase the strength, hardness and wear-resistance of the parent metal without necessarily having a deleterious effect on its ductility and formability. However, silver's high reflectivity and brilliance is adversely affected by alloying. Fortunately, additions of copper up to about 10 per cent increase strength and hardness considerably, and although the colour and lustre are changed slightly the appearance is still pleasing. Furthermore, the colour can be restored by subsequently electroplating articles with fine silver, and this is common practice in the silverware trade.

Silver–copper alloys

In the United Kingdom two legal standards (as defined by the Hallmarking Act 1973) exist for silver for use in silverware and jewellery. These are the Britannia Standard of a minimum fineness of 958.4 (95.84 per cent Ag) and Sterling Standard of minimum fineness 925 (92.5 per cent Ag). Since the Sterling Standard is more commonly used in the UK, as well as in most other countries of the world, it is also often referred to as Standard Silver. Although the Act does not say that the remainder should be copper, it usually is.

Alloys of a composition very similar to that of sterling silver were developed and used principally by the Romans, as analysis of items of Roman silver such as the Mildenhall Treasure in the British Museum has shown. However, the incorporation of the word 'sterling' into the English language occurred during the 12th century AD. Henry II brought over expert melters and coiners from the Easterling region in East Germany to improve and standardize coinage which was becoming debased, and since that time the sterling standard has been recognized.

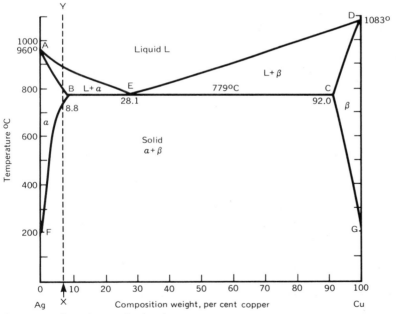

Figure 6.1. Phase diagram for the silver-copper system

In order to understand the effect of adding copper to silver, let us look at the binary equilibrium diagram or phase diagram for the Ag-Cu system (Figure 6.1). The diagram shows a vertical axis plotted as a temperature scale and a horizontal axis plotted in terms of per cent alloy composition such that the left-hand bottom corner represents fine silver and the vertical line drawn from it represents the effect of increasing temperature on the structure of silver. Similarly, the bottom right-hand corner represents pure or fine copper. The uppermost line across the diagram shows the temperature at which any alloy composition becomes fully molten and is known as the 'liquidus' line. You will see that it decreases from the melting point of fine silver at 960.5°C to a minimum of 779°C as copper additions are increased to 28.1 per cent and then increases to 1083°C, the melting point of pure copper, as the copper content is further increased.

Whereas pure metals melt and solidify at one characteristic temperature, i.e. their melting point, most alloys melt and solidify over a range of temperature. The line ABECD represents the temperature at which an alloy will be completely solid on cooling. This line is referred to as the 'solidus'. The liquidus and solidus lines merge at one particular composition and at a

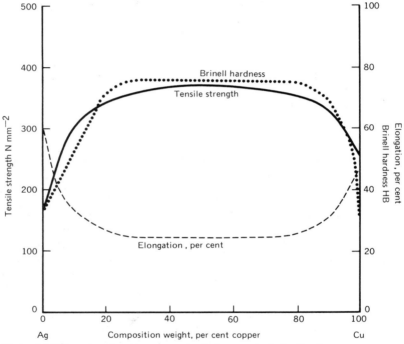

Figure 6.2. Effect of composition on mechanical properties in Ag-Cu alloys

minimum temperature in this diagram (71.9 per cent Ag–28.1 per cent Cu and 779°C). This is known as a eutectic composition and eutectic temperature from the Greek word 'eutektos' meaning 'easily melted'. Not all alloy systems show eutectic behaviour. Alloys in the phase fields bounded by the solidus and liquidus will be partially melted.

If copper is gradually added to silver, some silver atoms in its face-centred cubic lattice (Chapter 3) are replaced by copper atoms. We call this a solid solution of copper in silver. Because the size of a copper atom is different from that of a silver atom, distortion of the lattice occurs, and there is a limit to the amount of distortion the lattice can withstand. This limit is very small at room temperature but increases with temperature to a maximum of 8.8

per cent Cu at 779°C. The line FB in the diagram represents this limit of solid solubility. Alloy compositions to the left of the line FBA are single-phase solid solutions.

Similar behaviour is found at the copper-rich end of Figure 6.1 where the line CG represents the limit of solid solubility of silver in copper. Alloy compositions at temperatures in the region bounded by FBEDG are solid but consist of a mixture of the two phases on either side, in this case a mixture of crystals of silver-rich solid solution (α) and copper-rich solid solution (β). Greek letters are used to indicate the various single-phase fields in these diagrams.

It is the distortion of the lattice in solid solutions which in part leads to the increase in strength and hardness found on alloying. The greater the size mismatch, the greater the distortion and the greater is the solid-solution strengthening effect.

To summarize, the phase diagram conveniently represents alloying behaviour in terms of microstructure and melting and solidification behaviour for any composition and temperature in the system being studied. For a particular metal or alloy system, it is the microstructure which largely determines the mechanical properties, and this is demonstrated in Figure 6.2 for the Ag-Cu system.

Sterling silver

Let us now look specifically at sterling silver 92.5 per cent Ag–7.5 per cent Cu. A vertical line XY is drawn from this point on the composition axis in Figure 6.1. Following the line downwards from Y, we see that the alloy starts to solidify at about 910°C and that solidification is complete at about 810°C, giving a single-phase α solid solution. On further cooling, the line XY intersects the limit of solubility FB at 720°C. This means that crystals of the second phase β start to form in the α solid solution matrix, increasing in amount until at room temperature the structure consists of about 90 per cent of the α phase and 10 per cent β phase. It must be emphasized that the sequence just described is obtained only with slow equilibrium cooling conditions.

For the practising silversmith, sterling silver is used either in the cast state, i.e. a casting, or in the wrought condition, e.g. sheet for flatware and holloware, tube, rod and wire. Because solidification and cooling is fairly rapid during casting, solidification is not normally complete until the eutectic temperature of 779°C is reached and a non-equilibrium structure of crystals of α phase surrounded by a eutectic structure is obtained. The α phase crystals have a characteristic shape known as dendrites (Figure 6.3). The eutectic structure is in itself a finely-dispersed mixture of small α and β crystals.

With wrought products, the cast structure is destroyed by a combination of working and annealing. The final structure which is usually obtained after annealing at about 650°C followed by quenching into water or a pickle to remove oxidation products from the surface consists of α solid solution with perhaps a very small amount of β phase present as small sub-microscopic particles (Figure 6.4). This gives the alloy its softest and most ductile condition.

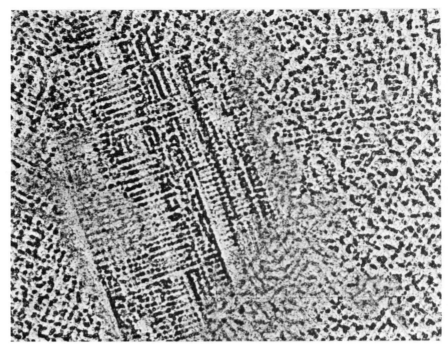

Figure 6.3. Microstructure of cast sterling silver (magnification ×200)

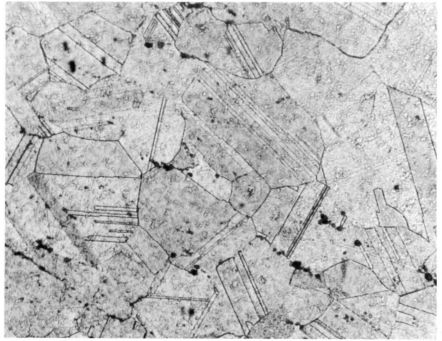

Figure 6.4. Photomicrograph of wrought and annealed sterling silver (magnification ×500)

It is not commonly known among silversmiths that wrought sterling silver can be considerably strengthened by applying a special heat treatment to the manufactured article. The treatment is known as age-hardening, and it comprises heating to 750°C for 30 minutes to obtain 100 per cent single phase α and then quenching rapidly to prevent the formation of any β phase. Subsequent ageing at 300°C for one hour allows a very fine dispersion of β phase particles to form in the α structure, imparting a considerable increase in strength and hardness. It is obvious that this treatment must be done using a temperature-controlled furnace. However, one word of warning must be given to those contemplating using an ageing treatment. It cannot be given to articles which have been silver-soldered, since 750°C is above the soldering temperature causing separation of the individual pieces making up the article. Nor can it be done prior to soldering, as heating to the soldering temperature will destroy the hardening effect given by the low-temperature ageing. Nevertheless, considerable improvements are gained with flatware and single-piece holloware.

Table 6.1 gives typical properties for sterling silver in the cast, annealed and aged conditions.

Table 6.1. Mechanical properties of silver alloys

Composition	Condition	Hardness HV	Tensile strength (N mm^{-2})	Elongation (%)	Erichsen Cup depth (mm)
Sterling silver	As cast	63			
	Annealed	56	300	45	10.2
	Cold-worked	140/180	550	4	
	Aged	110/120	350	16	
Britannia silver	Annealed	45			
800 Silver	Annealed	79	300	28	
Oxid^n Hard^d	Hardened	140	430		

For some purposes, such as spinning and chain-making, a more ductile grade of sterling silver is often preferred, and to achieve this some of the copper is replaced by cadmium (Cd) or more recently by zinc (Zn). Unfortunately, cadmium forms a highly toxic oxide which may be given off as brown fumes during melting and casting of cadmium-containing alloys, and unless adequate ventilation can be provided, zinc is preferred for safety reasons.

Britannia silver

In 1696, the standard of wrought plate in England was raised from 925 fineness to 958.4 because the conversion of silver coin into plate was interfering with the economy. So from 1696 to 1720 this was the only legal standard for silverware. Inevitably the lower copper content meant that a reduction in hardness and strength was obtained, and this was undesirable. Consequently, the sterling standard was re-introduced in 1720 and since that date the Britannia Standard has remained as an option. It is still used occasionally for silverware, particularly for reproduction pieces. An article made from Britannia silver may be recognized from the fact that the hallmark of the 'lion passant' is replaced by the figure of Britannia.

The term Britannia Silver must not be confused with Britannia metal, which is an alloy of 92 per cent tin–6 per cent antimony–2 per cent copper, and was used for cheaper classes of electroplate in the late 19th and early 20th centuries.

Other grades

Although only two standards are recognized for hallmarking purposes in the UK, this is not the case in other countries. For example, in Germany there are lower grades of fineness, notably 800 grade which is marked accordingly and which is widely used for tableware because of its increased hardness and strength. The composition is 80 per cent Ag–20 per cent Cu. However, the relatively high copper content imparts a definite yellowish tinge to the colour, and it is necessary to subsequently electroplate with fine silver. Other European countries also use 800 silver.

Silver-zinc alloys containing about 90 per cent Ag are produced in India for the manufacture of silverware. Indian silver has a characteristic whitish appearance resulting from the use of zinc instead of copper as the alloying addition.

Recently in the UK a grade of wrought silver sheet and wire has been introduced, known as 'oxidation-hardening silver'. This alloy is almost fine silver with controlled additions of less than 0.5 per cent of magnesium (Mg) and nickel (Ni). It is soft and ductile and can be easily formed. Once the item of silverware has been made, it is hardened simply by placing it in a furnace at 800°C for 1–8 hours, depending on section thickness. Oxygen from the surrounding air diffuses into the sheet or wire where it reacts with the magnesium to form hard submicroscopic particles of magnesium oxide (MgO) giving a hardness of 140–145 HV, which is much higher than that of aged sterling silver. Furthermore, soldering operations can be done on hardened pieces since it is impossible to re-soften the hardened metal. Another advantage is that as no copper is present, the alloy is not susceptible to firestain, a phenomenon which is discussed fully in Chapter 8. Care must be taken to avoid introducing oxygen at all stages of manufacture including annealing until the final hardening operation. Since great care has to be taken in the production of the alloy in the form of sheet or wire, and also because the silver content is higher, it is more costly than sterling silver and to some extent this has precluded its use on a wider scale. Articles made from oxidation hardening silver may be hallmarked as Britannia Silver although as stated they contain about 99.5 per cent Ag.

Tarnishing of silver and its alloys

It is well known that when silver is exposed to the atmosphere it tarnishes. Initially, the surface becomes covered with yellowish-brown flecking which gradually becomes a continuous film and darkens until it becomes black. At this stage, the tarnish film is quite thick and considerable effort has to be expended in removing it and restoring the surface to its original brilliant lustre. The film is silver sulphide (Ag_2S) and it is formed principally by the reaction between hydrogen sulphide gas (H_2S) and silver in the presence of oxygen from the air or moisture:

$$2Ag + H_2S + \frac{1}{2}O_2 = Ag_2S + H_2O$$

Sulphur-dioxide gas (SO_2) also has a role in tarnish formation. Unfortunately, these gases are present to some extent in the atmosphere due to pollution from the combustion of sulphur-bearing substances, e.g. coal. Prior to the industrial revolution and the use of coal for domestic fires, tarnishing occurred less rapidly on silverware than it has in more recent times.

The presence of dust in the atmosphere accelerates the rate of tarnishing, as does exposure to sunlight. In this case, the heating effect of the sun increases the rate of reaction.

Contact with sulphur-bearing substances such as egg yolks, onions and vulcanized rubber also leads to tarnishing. Matt and textured surfaces tarnish more rapidly than highly polished surfaces due to an increased surface area.

Silver-copper alloys tarnish at a greater rate than fine silver. The tarnish film is more complex because both silver and copper sulphides are present and copper oxide has been detected also in the film.

Removal of tarnish

In spite of some of the tarnish-protection methods which are available and which are discussed in the following section, it is necessary to clean silver which is in daily use such as tableware. Tarnish removal is done either mechanically or chemically. Mechanical cleaning consists of abrasion with a finely-ground powder such as chalk or jeweller's rouge made up as a paste and spread on a suitable polishing cloth. This not only removes the tarnish film but polishes the silver surface. A number of proprietary pastes or impregnated cloths are available on the market, e.g. the firm of Goddards have a number of products which may be purchased at jewellers shops and hardware stores.

Chemical cleaning involves immersion in a solution which dissolves the tarnish film or reduces it to silver. Suitable solutions are sodium thiosulphate or ammonia in water. Some proprietary solutions contain long-chain molecule aliphatic compounds which surprisingly contain sulphur at the active end of the molecule. These active ends attach themselves to the cleaned silver surface to produce a thin transparent protective film which does not interfere with repolishing.

Protection against tarnishing

A considerable amount of time, money and effort has been spent over the last few decades to find means of protecting silver and its alloys against tarnishing.

One line of attack has been to replace copper in alloys by another metal in an attempt to increase tarnish resistance. Palladium, zinc and cadmium all impart some resistance but these alloys are not commercially viable because of certain restrictions such as

1. expense,
2. the required alloying addition exceeds the standard imposed by the Hallmarking Act,
3. the alloy must have the same lustre as that of silver,
4. the alloy must be easily manufactured,
5. a high degree of formability is required,
6. annealing and soldering procedures must be feasible.

As these difficulties have not yet been resolved, other methods of protection have been employed.

Plating

Obviously the metal used for plating onto silver must itself be tarnish-resistant and, because the attractive silver lustre is covered, it must itself have a high reflectivity throughout the visible light spectrum. The only metal which fits these requirements is rhodium which has a reflectivity of about 80 per cent. Because of its high cost, rhodium plating adds to the value of silverware.

Lacquers

Clear nitro-cellulose and other lacquers may be applied to silver by spraying or dipping. They have the advantage of being transparent, clear and resistant to light and heat. They are not suitable for articles which are in continual use, such as tableware, because they are easily scratched exposing the underlying surface to tarnish attack, but they do afford valuable protection to museum pieces which are rarely handled.

Anti-tarnish wrapping paper

Most proprietary wrapping papers contain chlorides from the bleaching process used in paper-making. In the presence of damp air these react with sulphur-containing substances to form hydrogen sulphide. This in turn atacks silver wrapped in such paper.

Anti-tarnish papers have been developed which inhibit tarnishing. These papers are impregnated with salts such as copper acetate which preferentially absorb hydrogen sulphide and prevent it attacking silver for up to two years.

Sealed polythene bags containing silverware are popular as they prevent ingress of air and provide protection. At the same time the silverware may be easily displayed and inspected. Cellophane bags, although cheaper, are often unsatisfactory due to tarnishing caused by adhesives used in their manufacture.

7
Alloys of gold: the carat golds

As with silver, it is common practice for bullion dealers and refiners to express the quality of gold in terms of fineness in parts per thousand. Pure gold has a fineness of 1000. In the UK the Hallmarking Act of 1973 recognizes four legal standards. These are 916.6 fine (22-carat); 750 fine (18-carat); 585 fine (14-carat); 375 fine (9-carat). You will see that the carat system is also used to describe these standards. The word 'carat' has arisen from an arabic word 'Kîrât' for the seed for the locust bean tree. Because the seeds have the characteristics of similar size and weight they were used by

Table 7.1. Caratages of gold

Caratage	Fineness	% gold
14	1000	100
22	916.6	91.66
20*	833.3	83.33
18	750	75
15	622	62.2
14	585	58.5
12	500	50
10	416	41.6
9	375	37.5
8	333	33.3

* Rarely used, but the horse-racing trophy, the Ascot Gold
  Cup, was made from this caratage.

African and Arab traders for weighing gold. Formerly, one gold coin or 'mark' weighed 24 carats but now it is taken to indicate how many twenty-fourth parts of the metal are gold. Hence, pure gold is 24-carat. 22-carat gold means that 22 out of 24 parts of the alloy are gold, i.e. 91.66 per cent Au; 14-carat is 14 out of 24 parts or 58.5 per cent Au, and so on. Incidentally, the carat is still used in gemmology to denote the weight of a gemstone, one metric carat having been standardized at 0.2 g.

Between 1854 and 1932, 15- and 12-carat standards were recognized in the UK but were replaced in 1932 when the 14-carat standard was introduced. Other carat (ct) gold standards (spelt karat, kt, in America and on the Continent) are recognized in other countries. For example, the lowest in the USA is 10 kt (416 fine) and in Germany it is 8 kt. Table 7.1 gives the commonly-accepted caratages in terms of fineness and percentage of gold content.

Pure gold is much too soft to be used for jewellery manufacture and its main use has been for coinage and various industrial purposes. To a large extent this is also true for the 22 ct golds although they have traditionally been used for wedding rings. Generally, high-class jewellery is made from 18 ct golds because they have excellent tarnish and corrosion resistance coupled with good strength and ductility imparted by the 25 per cent of alloy additions. Furthermore, the colour of the 18 ct and lower-caratage golds is profoundly affected by the alloy content as we shall see in the next section.

The tarnish and corrosion resistance decreases with decreasing gold content so that the use of the relatively cheaper but popular 9 ct golds does give rise to certain problems such as tarnishing and stress-corrosion cracking.

Alloying behaviour

In order to appreciate the development and use of the carat-gold alloys it is necessary to examine briefly the alloying behaviour of gold. To do this it is convenient to divide the carat golds into two classes, namely, the coloured carat golds and the white carat golds.

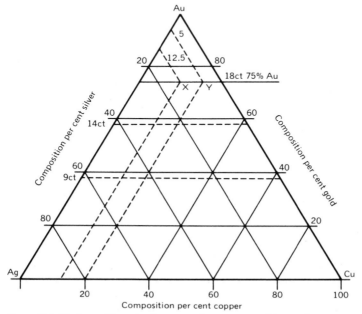

Figure 7.1. Horizontal base for the Au-Ag-Cu ternary equilibrium diagram

Coloured carat golds

The coloured carat golds are based on the gold-silver-copper (Au–Ag–Cu) alloy system although other additions, notably zinc (Zn), may be present in the lower-caratage alloys. In an 18 ct alloy the remaining alloy content of 25 per cent will consist of silver and copper in varying proportions. Such a system is referred to as a ternary alloy system and its alloying behaviour is represented by a ternary equilibrium diagram. Ideally this should be drawn in three dimensions with temperature again as the vertical axis and the compositions of the three binary systems, in this case, Au–Ag, Au–Cu and Ag–Cu, which make up the ternary system plotted on a horizontal base around the three sides of an equilateral triangle. Figure 7.1 shows this

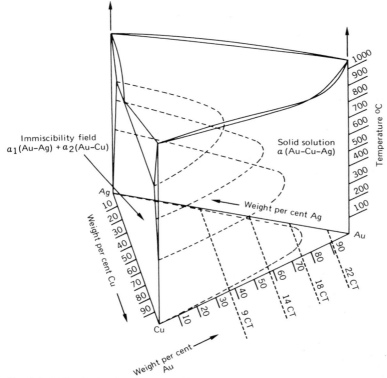

Figure 7.2. The Au-Ag-Cu ternary equilibrium diagram (the formation of AuCu and AuCu$_3$ have been omitted for clarity)

horizontal base on which, for example, all 18 ct Au–Ag–Cu alloys have compositions along a line drawn from 75 per cent Au–25 per cent Ag to 75 per cent Au–25 per cent Cu. An alloy containing 75 per cent Au–12.5 per cent Ag–12.5 per cent Cu will be positioned midway along this line (point X) while an alloy containing 75 per cent Au–20 per cent Ag–5 per cent Cu will be positioned at point Y which is four-fifths of the way along the line towards the Au–Cu side of the diagram. Similar arguments can be made for the other carat-gold alloys. Note that the composition lines for 22, 14 and 9 ct alloys are also drawn.

The complete three-dimensional diagram is depicted in Figure 7.2 and the bounding binary systems for Au–Ag in Figure 7.3 and Au–Cu in Figure 7.4. The Ag–Cu diagram was shown in the previous chapter (Figure 6.1). It is common practice to draw horizontal sections at temperatures of interest to discover the phases that may be present in the microstructure.

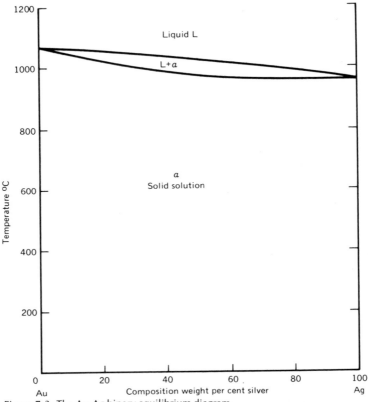

Figure 7.3. The Au-Ag binary equilibrium diagram

Horizontal projections can be used to show the liquidus temperatures of the alloys in the system, i.e. the temperatures at which they start to solidify on cooling from the molten state (Figure 7.5), or to show how the colours of the carat golds change with alloy composition (Figure 7.6). The latter diagram is known as *the colour triangle*.

The colour triangle shows that the colour of the carat gold depends markedly on the varying proportions of silver and copper in the alloy as well as on the caratage. Hence, we have the red golds in which the alloying addition is predominantly copper; green golds, which really do have a pronounced greenish tinge and in which silver is the major addition; and the yellow golds where the silver and copper are present in similar amounts. Consequently, it is easy to see how items of jewellery made of a certain caratage can display different colours. Recently a system of standardizing the colours of carat golds was introduced in Switzerland, Germany and France.

Zinc additions up to about 10 per cent are made to 9 and 14 ct alloys partly to offset the reddish hue imparted by copper thereby giving a rich

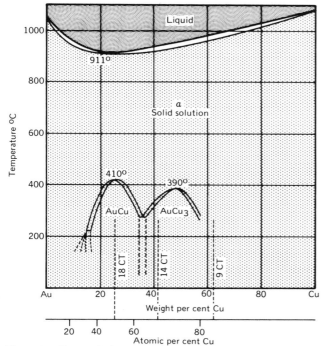

Figure 7.4. The Au-Cu binary equilibrium diagram

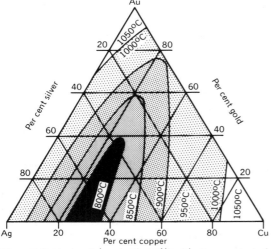

Figure 7.5. Horizontal projection of liquidus temperatures for Au-Ag-Cu alloys

yellow colour and partly to improve castability and mechanical properties. Cadmium may be added to make pale-green golds.

Other interesting colour effects have been found. An intense violet colour is obtained in a 75.8 per cent Au–24.2 per cent aluminium (Al) alloy. Unfortunately, this composition corresponds to that of an intermetallic compound $AuAl_2$ which is very brittle and cannot be formed into items of jewellery, although its use for decorative gold surfaces incorporating particles of $AuAl_2$ has been reported. A very attractive ice-blue colour can be obtained with about 54 per cent indium (In), Au_2In; olive green with 4.72 per cent potassium (K), Au_4K, and violet with 9 per cent K, Au_2K. Again, these are all brittle intermetallic compounds.

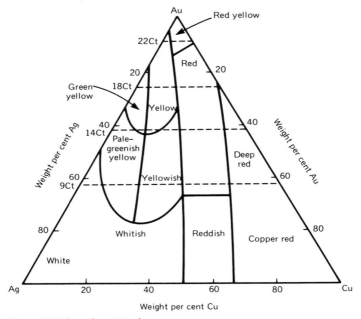

Figure 7.6. The colour triangle

The mechanical properties of the coloured carat golds are dependent on the various proportions of alloying additions which are present. As seen in Chapter 6, addition of solute atoms leads to a distortion of the parent metal lattice giving a solid solution strengthening effect. The amount of lattice distortion and hence strengthening is dependent on the relative sizes of the parent and solute atoms and also on the amount of solute added. The sizes, expressed as atomic radii, of gold, silver and copper are as follows:

Gold	0.1442 nm
Silver	0.1443 nm
Copper	0.1277 nm

Since the sizes of gold and silver atoms are similar, one would not expect a great strengthening effect with silver additions since lattice distortion is minimal, whereas copper should impart considerable strengthening and this is indeed the case.

Silver atoms can completely replace the gold atoms from 0 to 100 per cent in the FCC lattice because of the similarity in size, i.e. Au–Ag alloys form a complete series of solid solutions (Figure 7.3). The situation is more complex when copper is also added. Binary Au–Cu alloys (Figure 7.4) form a complete series of solid solutions above 410°C but in the composition range from 15 per cent Cu to 60 per cent Cu intermetallic compounds AuCu and AuCu$_3$ are found below 410°C. Rapid cooling from above 410°C will prevent the formation of these compounds, which are referred to as 'ordered phases', thereby retaining the relatively weaker 'disordered' solid solution. The presence of the ordered phases which can be produced by ageing between 150–400°C leads to considerable hardening (Figure 7.7). It should be appreciated that an 18 ct red gold falls in the middle of the range centred on AuCu.

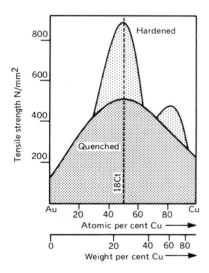

Figure 7.7. Effect of ordering on the strength of Au–Cu alloys

In addition, in the ternary system there is a limit to the combined number of both silver and copper atoms which can be dissolved in the gold lattice, and in certain composition ranges and temperatures the structure of the alloys consists of two crystal phases, one based on the Au–Cu rich solid solution and the other on the Au–Ag rich solid solution, i.e. the two-phase region extends into the ternary equilibrium diagram. This is shown in Figure 7.8 which shows the extent of the two-phase region at different temperatures. A consideration of the horizontal lines depicting the compositions of the standard carat golds shows that all 22 ct coloured alloys are single phase at all temperatures, as are most 18 ct golds, but both 9 ct and 14 ct alloys show very limited solid solubility at lower temperatures. As with sterling silver, many 9 ct and 14 ct alloys can be age-hardened by quenching from high temperatures, say 650°C, to retain the solid solution and then heating at 300°C to precipitate out the second phase which leads to appreciable strengthening (Figure 7.9).

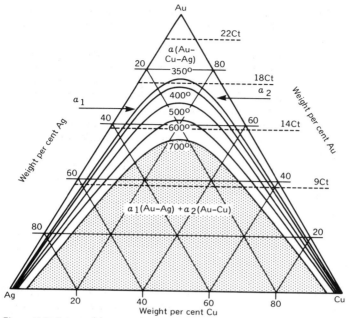

Figure 7.8. Extent of the two-phase region at different temperatures

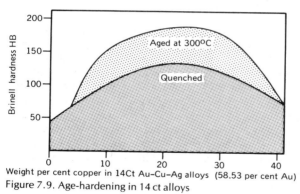

Figure 7.9. Age-hardening in 14 ct alloys

It is for these reasons that goldsmiths and jewellers are advised to quench after annealing to achieve maximum softness and formability as slow cooling leads to the formation of ordered phases and/or precipitate phases with consequent hardening and loss of formability.

Table 7.2 gives an indication of the mechanical properties of the coloured carat golds in both the annealed and aged conditions.

White carat golds

The colour triangle shows that it is possible to produce a white 9 ct gold in the Au–Ag–Cu system with up to 6 per cent Cu such that the Ag + Cu content is 62.5 per cent. Such alloys are used for jewellery manufacture but they are rather unsatisfactory as they are too soft. Furthermore, the tarnish resistance is poor because of the high silver and copper content. High-carat white golds cannot be made using this system.

The commercial white golds were originally developed early this century as substitutes for platinum which was then much more expensive than gold. The requirements were good tarnish resistance, good formability and suitability as a mounting medium for gemstones. Only palladium, nickel and to a lesser extent manganese (Mn) give a significant whitening effect coupled with the above requirements, and of these manganese has the least effect. Platinum also whitens gold but palladium is considerably more effective and cheaper.

There are three classes of white carat golds: (a) the palladium white golds; (b) the nickel white golds; (c) white golds with both palladium and nickel; but it should be noted that small additions of copper, nickel, zinc, manganese, platinum and silver may be present in alloys from all classes.

(a) *The palladium white golds*. These can be produced at all caratages up to 20 ct. A 22 ct white gold cannot be made because there is insufficient alloy content to give the desired whitening effect. The alloys are characterized by high melting ranges, typically 1100 to 1320°C depending on composition, and a high degree of softness and malleability. Because the Au–Pd system exhibits a complete range of solid solubility at all temperatures they cannot be hardened by heat treatment unless copper is also present. They can, of course, be hardened by cold-working.

(b) *The nickel white golds*. These are also obtainable up to 20 ct. They are characterized by much lower melting ranges than those of the palladium white golds, typically 800 to 1050°C. Nickel has a large hardening effect on gold, and considerable increases in hardness, tensile strength and limit of proportionality are obtained in these alloys in both the cold-worked and annealed conditions. The high limit of proportionality means that a high degree of springiness is achieved which makes them eminently suitable for use as spring clips, hinge pins, etc. However, this same property makes the alloys unsuitable for gemstone setting. The palladium white golds have to be used for white gold stone settings and mounts. Although the ductility is decreased, the alloys are not brittle and they can be fabricated into the same forms as those manufactured from the coloured carat golds.

Just below the solidus the alloys are single-phase solid solutions but on slow cooling or ageing heat treatments they separate out into gold-rich and nickel-rich phases. Consequently, they can be age-hardened, but this is not usually desirable as phase separation leads to (1) loss of ductility and workability, (2) a loss in whiteness due to the appearance of the yellow gold-rich phase, and (3) a possible reduction in corrosion resistance.

(c) *The palladium-nickel white golds*. More recently, a class of 18-carat white golds containing both palladium and nickel have become commercially available. Although more costly than the nickel white golds (because of the palladium content), they have improved corrosion resistance and colour and have adequate ductility for most purposes. Gemstone setting may be possible for some compositions.

Table 7.2. Mechanical properties of the coloured carat golds

Type	Composition %			Colour	Solidus °C	Condition	Hardness		Tensile strength N mm⁻²	Elongation %
	Ag	Cu	Zn				HB	HV		
22 ct 91.6% Au	8.4	–	–	Yellow	1024	Annealed	30		157	41
						Cold-worked	74		283	
	5.5	2.8	–	Yellow	995	Annealed		52	220	27
						Cold-worked		138	440	0.5
	4.2	4.2	–	Deep yellow	980	Annealed	57		290	35
						Cold-worked	123		503	
	–	8.4	–	Red	930	Annealed	66		373	41
						Cold-worked	155		610	
18 ct 75% Au	25	–	–	Green yellow	1028	Annealed	32		185	36
						Cold-worked	93		337	
	16	9	–	Pale yellow	895	Annealed		135	500	35
						Cold-worked		210	800	1
	12.5	12.5	–	Yellow	885	Annealed		150	520	45
						Cold-worked		225	900	1.5
						Aged		230	750	15
	9	16	–	Pink	880	Annealed		160	550	40
						Cold-worked		240	920	2
						Aged		285	850	7
	4.5	20.5	–	Red	890	Annealed		165	550	40
						Cold-worked		240	950	1.5
						Aged		325	950	4

	–	25	–	Deep red	910	Annealed		165	514	42
						Cold-worked		240	873	
						Aged		340		
14 ct 58.5% Au	41.5	–	–	Pale green	920	Annealed	34		198	38
						Cold-worked	97		347	
	35.5	6	–	Green	841	Annealed	78		441	35
						Cold-worked	139		778	
	20.5	21	–	Yellow	830	Annealed		190	580	25
						Cold-worked		260	1000	1.5
						Aged		270	800	3
	10.3	27.7	3.8	Pink	810	Annealed		148	440	35
						Cold-worked		288	1079	1
						Aged		242	593	3
	9	32.5	–	Red	850	Annealed		160	550	45
						Cold-worked		270	1000	1.5
						Aged		260	700	12
	–	41.5	–	Deep red	907	Annealed	83		398	52
						Cold-worked	155		821	
9 ct 37.5% Au	62.5	–	–	Whitish	950	Annealed	34		196	39
	55	7.5	–	Pale yellow		Annealed	80		397	32
	42.5	20	–	Yellow	790	Annealed	120		494	25
	31.25	31.25	–	Rich yellow	830	Annealed	110		540	25
	20	42.5	–	Pink		Annealed	110		340	30
	7,5	55	–	Red		Annealed	90		448	40
	–	62.5	–	Deep red	970	Annealed	70		417	37

Much of the data shown in Tables 7.2 and 7.3 has been taken from the Gold Alloy Data published by the International Gold Corporation in *Aurum*.

Table 7.3. Mechanical properties of the white carat golds

Type	Composition %	Solidus °C	Condition	Hardness HB	Hardness HV	Tensile strength N mm⁻²	Elongation %
Pd white							
18 ct 75% Au	10% Pd 10.5% Ag 4.5% (Cu + Ni + Zn)	1020	Annealed		95	379	33
			Cold-worked		216	703	1
14 ct 58.5% Au	10% Pd 28.5% Ag 2.5% (Cu + Ni + Zn)	1015	Annealed		82	414	24
			Cold-worked		195	653	0
Ni white							
18 ct 75% Au	14% Ni 11% (Cu + Zn)	910	Annealed		220	710	42
			Cold-worked		350	1235	1.6
14 ct 58.5% Au	14.5% Ni 20% Cu 7% Zn		Annealed	152		726	42.8
	15.3% Ni 25.8% Cu 0.4% Zn		Annealed	177		698	51.3
9 ct 37.5% Au	17.5% Ni 27.6% Cu 17.4% Zn					685	35
Pd–Ni white							
18 ct 75% Au	15% Pd 10% (Cu + Ni)	1095	Annealed		180	550	33
			Cold-worked		280	940	3.5

Certain precautions must be taken when casting, working and annealing the nickel white golds, and these will be detailed in later chapters.

The mechanical properties of typical white golds are given in Table 7.3 and these may be compared with those from Table 7.2 for the coloured golds.

Specific characteristics of the carat golds

Having reviewed in general terms the alloying behaviour, it is pertinent to look at certain characteristics of the alloys comprising the four legal standards adopted by the United Kingdom and many other countries. It is obviously impossible in a book of this nature to give a detailed summary of all the alloys for each class, and indeed it would not be desirable to do so.

Every manufacturer and craftsman should select the alloy most suitable for a particular application paying due attention to the aspects of colour, required strength and wear resistance, formability, amenability to hardening by heat treatment after fabrication, solderability, cost, etc. Information on these aspects is readily given by the bullion dealers and manufacturers of carat gold alloys in semi-fabricated form.

22-carat golds

Gold containing 91.66 per cent Au and 8.34 per cent of alloying additions has also been referred to as 'standard gold' because of its use for British gold coinage. It was first made the legal standard for this purpose in 1527 but it was not until 1575 that it was legalized for the manufacture of gold wares.

The colour ranges from pale yellow when the addition is silver to red-yellow when the addition is copper. As mentioned earlier, 22 ct is used almost exclusively for wedding rings and a composition of about 6 per cent Cu and 2 per cent Ag is preferred as it has reasonable strength and ductility. The alloy cannot be hardened by heat treatment.

18-carat golds

These alloys contain 75 per cent gold and 25 per cent alloy additions and are the most highly-favoured for high-class jewellery manufacture because of

1. the range of colour that can be produced,
2. the excellent mechanical properties, and
3. excellent tarnish resistance.

The standard was first legalized by a statute of 1477, discontinued when the 22 ct standard was introduced in 1576 and then reintroduced in 1798.

The mechanical properties vary considerably with composition. In the coloured golds, increasing the copper content increases strength and hardness whereas a high silver content gives lower strength and hardness. A good general-purpose 18 ct has roughly equal amounts of copper and silver, i.e. 12.5 per cent of each, as this gives an alloy with a good yellow colour suitable for most classes of work. The high-copper 18 ct red golds are hardened greatly by air cooling from above 410°C with consequent loss in ductility due to ordering and precipitation effects.

18 ct nickel white golds contain copper and zinc in addition to nickel to improve workability, although some colour variation is experienced depending on actual composition. The 18 ct palladium white golds contain a certain amount of silver in addition to palladium and are relatively soft and easy to work. They are more expensive than the nickel white golds because of the cost of palladium. It has been mentioned that 18 ct white golds are produced which contain significant amounts of both nickel and palladium together with copper, zinc and silver.

14-carat golds

These alloys contain 58.5 per cent gold and 41.5 per cent alloy content. The standard was legalized in 1932 when the 15 ct and 12 ct standards were abolished. However, the use of 14 ct gold for goldware and jewellery can be traced back to the 14th century. It has found wide application for fountain-pen nibs tipped with a hard wear-resistant iridium alloy and for pencil cases and spectacle frames.

The melting ranges of the coloured 14 ct golds are in general a little lower than those of the 18 ct golds but the annealing temperatures are of the order of 100°C higher, 650°C compared with 550°C. This is because the two-phase region is extensive at the 14 ct compositions. In fact, alloys having a composition in the middle of the range, i.e. equal amounts of silver and copper, are very difficult to fabricate due to a tendency to crack during rolling or drawing. However, the ductility of these alloys can be increased by small additions of nickel and zinc. It is advisable to anneal and quench all 14 ct alloys prior to working. All 14 ct coloured golds containing more than 5 per cent copper can be hardened by quenching from 650°C followed by ageing at about 300°C.

The tarnish and corrosion resistance, particularly for the 14 ct red golds, is not as good as that of the 18 ct alloys.

9-carat golds

These alloys contain 37.5 per cent gold and 62.5 per cent alloy addition. The standard was legalized in 1854 and immediately became popular for the cheaper classes of jewellery. Hence, they have attained a high degree of importance in the jewellery industry, particularly in the United Kingdom where the greatest proportion of all gold jewellery submitted for hallmarking is 9 ct gold.

The commercial alloys are rarely simple Au–Ag–Cu ternary alloys as they contain appreciable quantities of zinc which tends to reduce their melting ranges but counteracts the reddening effect of the high copper content. As might be expected, the ranges of colour and mechanical properties which can be achieved by variation in composition is very large. Some of these alloys are amenable to precipitation hardening by quenching from 650°C and ageing at 300°C for ten minutes.

In reality, with such a high alloy content, particularly of the base metals, the 9 ct golds can hardly be considered as noble metal alloys. Their tarnish and corrosion resistance is very inferior to that of the 14 and 18 ct alloys.

Tarnishing behaviour and chemical properties

It is well known that silver and its alloys tarnish but it is not always appreciated that the lower-caratage gold alloys also tarnish. Sometimes customers complain to jewellers that skin and clothing have been blackened by their gold jewellery. Although the jewellery itself may not actually look tarnished, it is likely that traces of tarnish film have in fact rubbed off and caused the observed blackening.

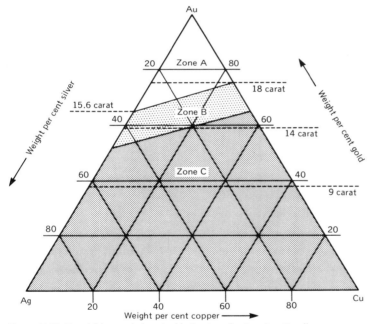

Figure 7.10. Tarnishing and chemical behaviour for Au–Ag–Cu alloys

It is not really surprising that many of the carat golds tarnish, given that the base metal and silver contents may be quite high and furthermore, in alloys which have two or more phases present in their microstructure, one of these often contains a high proportion of base metal. Sulphur-bearing gases cause tarnishing, as do human perspiration (some people may be more affected than others), soaps and cosmetics.

The immunity to attack by acids also depends on alloy composition. In 1921, a German metallurgist, Professor Tammann, conducted a systematic study of the tarnishing and chemical attack by acids of Au–Ag–Cu ternary alloys. He found that provided the gold atom/total atom ratio in the alloy exceeded 50 per cent, the alloys were immune to tarnishing and are only soluble in aqua regia, cyanide solutions and, under special conditions, chloride solutions. Figure 7.10 shows that alloys in this category are located above a line drawn from the 64.6 per cent Au–Ag alloy to the 75.6 per cent Au–Cu alloy, i.e. fractionally above the 18 ct deep-red gold composition so that all 22 ct and practically all 18 ct golds are immune from tarnishing and strong acid attack.

A second zone exists with the lower boundary line running from 52.3 per cent Au–47.7 per cent Ag copper-free green gold to 65 per cent Au–35 per cent Cu silver-free red gold. Strong acids dissolve out the silver and copper atoms from the surface leaving it enriched in gold, and attack then ceases.

Below this line the alloys disintegrate in strong acids and tarnish. This zone includes all 9 ct alloys and the 14 ct red golds. The presence of other metals such as zinc modify these compositional boundaries. Because tarnishing is obviously undesirable, jewellery made from susceptible alloys is often plated with a layer of 18 ct or 22 ct gold.

Blackening of skin and clothing has been experienced with high-carat gold jewellery. From what has been said above, this cannot be due to tarnish film removal. In this case, mechanical abrasion may occur whereby fine metallic particles are transferred from the surface of the article onto skin or clothing. Unfortunately, it seems that synthetic fibres used in modern textiles enhance this effect and again the only way to minimize it is to plate with hard gold.

Certain 14 and 9 ct golds are also subject to another form of corrosion known as stress corrosion cracking, but as this is influenced by working and annealing behaviour, discussion of this is best left to Chapter 10.

8

The platinum-group metals and their alloys

One important distinction exists between the use of platinum and the use of silver and gold for jewellery and decorative metalware. About 50 per cent of gold production is used for jewellery and the figure for silver is 17 per cent, whereas up to 1970 only about 7 per cent of the world's production of platinum was utilized for jewellery and the arts. However, during the last decade there has been an upsurge in the jewellery market for platinum, particularly in Japan, West Germany, the USA and the UK. Nevertheless, platinum-group metals are mainly used for industrial purposes and these will be reviewed in Chapter 15. In addition, although there is a wealth of information on the development of silver and gold alloys for the decorative arts, very little has been published in this field for the platinum-group metals and it is difficult to present a coherent picture for the jewellery alloys.

When we looked at the history of platinum in Chapter 2, mention was made of the work of William Wollaston. Between 1800 and 1820 he produced some 38 000 oz of refined platinum which were fabricated by the powder metallurgical techniques of compaction of the metal powder in a die, heating to a temperature below the melting point but sufficiently high to effect bonding between the powder particles (a process known as 'sintering'), and finally hot forging to consolidate the metal to full density. The main outlet was in the manufacture of boilers for the production of sulphuric acid. The next important person to appear on the scene was Percival Norton Johnson who set up a refinery in Hatton Garden, London, for the production of malleable platinum and palladium. In addition, iridium was produced for tipping gold pen nibs, an ideal material because of its hardness and wear resistance.

During the first half of the 19th century, other famous scientists were finding uses for platinum. Michael Faraday discovered the advantages of melting optical-quality glass in platinum crucibles, Humphrey Davy showed that platinum could be used to catalyse chemical reactions and Alfred Smee, Surgeon to the Bank of England, used platinum electrodes in his valuable work on electrodeposition of metals.

George Matthey, who had joined Johnson's firm as an apprentice at the age of 13 years, was largely responsible for the growth of the platinum metals industry in the second half of the 19th century, and is particularly known for developing the platinum–10 per cent iridium alloy for imperial and metric standards of weight and length. Mattey rose to become head of the firm founded by Johnson and the two names were joined to form the world-renowned firm of Johnson Matthey.

In this period, the problem of melting platinum on an industrial scale had been overcome by Deville and Debray with the introduction of lime block furnaces, lime and magnesium crucibles and oxy-hydrogen flames. These were used up to about 1918 when induction melting equipment became available.

Jewellery and decorative metalware

The platinum standard of 950 fineness (95 per cent minimum) has been accepted in the UK for hallmarking purposes, and this standard is likely to be adopted in many other countries.

Pure platinum is too soft for use in jewellery and so it is hardened by alloying. Among the early alloys developed for this purpose and which are still in use today are platinum (Pt)–4.5% copper (Cu), platinum–5% iridium (Ir), platinum–4.5% palladium (Pd), platinum–5% ruthenium (Ru) and platinum–4.5% cobalt (Co). Hardnesses in the annealed condition are shown in Table 8.1.

Table 8.1

Alloy	Hardness HV
Pt–4.5% Cu	80
Pt–5% Ir	110
Pt–5% Ru	130
Pt–4.5% Pd	65
Pt–4.5% Co	135

The copper content in the first alloy is insufficient to cause blackening due to oxidation on heating. The alloy can be used in both the wrought and cast state. The softer alloys are suitable for ornamental work such as mounts and gemstones. The harder alloys are not usually recommended for castings but they are suitable for wrought work, particularly for springs and clasps.

Recently, a new class of alloys has been developed with compositions in the range 95 per cent platinum–1.5 to 3.5 per cent gallium (Ga)–balance gold (Au) or indium (In). The Pt–Ga–Au alloys are particularly recommended for lost wax investment castings while the Pt–Ga–In alloys are used for springs and clasps.

Palladium is also important to the jewellery trade but mainly as an alloying constituent in the palladium white golds, as we saw in the previous chapter. A certain amount of palladium jewellery has been manufactured using alloys such as 60 per cent Pd–40 per cent Ag, 95 per cent Pd–4 per cent Ru–1 per cent Rh and 95 per cent Pd–5 per cent Ni, but this type of jewellery has now largely lost favour with the public.

Iridium has very little ductility and has to be hot-worked. Consequently it cannot be used in its own right as a jewellery metal. It is often found in combination with osmium, either as iridosmium when iridium predominates or osmiridium when osmium predominates. Its high hardness, wear and corrosion resistance make it suitable for tipping fountain-pen nibs.

Apart from its use as an alloying addition in some jewellery alloys, rhodium is favoured as an electroplated coating for silverware because of its relatively high reflectivity and tarnish resistance.

Ruthenium and osmium are virtually unworkable in the cast state and are processed by powder metallurgical techniques. Ruthenium is used for Pt-Ru jewellery alloys.

The platinum-group metals, with the exception of platinum itself, slightly discolour on heating at low temperatures due to the formation of thin oxide films but at certain critical temperatures, depending on the particular metal, the oxides volatilize and clean surfaces are obtained. It should be noted that oxide fumes from osmium are highly poisonous.

Lustre ware

One important application of platinum developed in the 17th century was the platinum decoration of porcelain and English lustre ware. Dissolving platinum in aqua regia and adding ammonium chloride gives a precipitate of ammonium chloroplatinate. Heating the chloroplatinate drives off the ammonium chloride leaving the platinum as a finely-divided powder. This is mixed with a flux and 'oil of spike lavender' and applied with a brush to the surface of the ceramic ware. Firing in the kiln imparts a silver-grey glaze.

9
Melting, alloying and casting

'When you first cast gold examine it by careful scraping and scratching around it in case there happens to be any air bubble or crack in it. This often happens when through carelessness, or negligence, or ignorance, or lack of skill of the founder (caster), it is cast either too hot or too cold, or too quickly or too slowly. If when you have cast it with care and circumspection you perceive a flaw of this kind in it, carefully dig it out if you can with a suitable tool. But if the bubble or crack is so deep that you cannot dig it out, you must cast again until it is sound.'

These are the words of Theophilus, a German monk and goldsmith, living in the first half of the 12th century. They are as true today as when they were written.

Solidification behaviour

We saw in Chapter 3 that, in the molten state, atoms are loosely held and move about relatively easily whereas, in the solid state, they are held in a regular geometrical array — the face-centred cubic structure for all the precious metals with the exception of ruthenium and osmium.

When solidification starts it does so from a few isolated points in the melt, usually from impurity particles or from the mould or crucible wall in much the same way as ice begins to form on a pond. This process is known as nucleation and it occurs when a few atoms come together to form tiny seed crystals at the melting (freezing) point of a metal or just below it and for alloys at the liquidus temperature where solidification occurs over a range of temperature. As more and more metal atoms are deposited on these seeds or nuclei they begin to grow as crystals having the geometrical lattice arrangement.

Imagine a nucleus as being a tiny cube floating in the melt and growing preferentially in the directions of the cube edges by the formation of primary spikes extending from all six faces of the cube into the liquid metal. Similarly, secondary and tertiary spikes branch out sideways from the primary arms, and eventually a tree-like metal crystal called a dendrite is produced (Figure 9.1).

Grain structure

In practice, a large number of nuclei are formed. All of them are capable of growth but the directions in which the dendrite arms grow in the melt depend on the orientation of each nucleus. Consequently, when the dendrites growing from different nuclei come into contact with each other, a boundary is formed at the contact interface. Metallurgists call this a grain boundary, i.e. the boundary between two crystals or grains. At this stage some liquid will still be trapped between the dendrite arms but as solidification proceeds, the arms grow thicker by metal deposition until

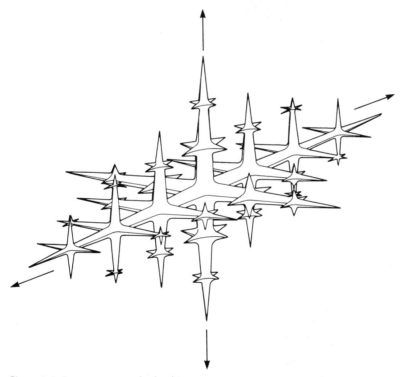

Figure 9.1. Representation of a dendrite

eventually all the intervening space is filled with solid metal. The sequence is shown in Figure 9.2. A good representation of a polycrystalline grain structure is obtained when a soap solution is shaken to produce a three-dimensional network of soap film boundaries. This is analogous to the grain boundary network in the metal.

Metallurgists examine the grain structure of metals by taking a cross-section of the cast metal, grinding and polishing the surface produced to remove the worked layer left by the cutting operation and then etching this surface in a suitable chemical reagent which preferentially attacks the metal at the grain boundaries.

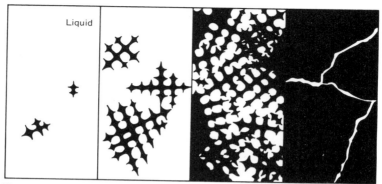

Figure 9.2. Nucleation and growth of dendrites to form a polycrystalline grain structure

Examination of the etched surface under a reflecting-light microscope will reveal the grain structure (Figure 9.3). In pure metals, there is normally no evidence of dendritic growth and only a mass of grains with boundaries between them are seen in the microscope. However, if the residual liquid metal is decanted during the solidification process or if a void remains in the centre of the casting, the dendrite arms may become visible.

Because each nucleus develops into a grain, the final grain size is determined by the number of nuclei. With a slow cooling rate from the melt only a few nuclei are formed and these can grow relatively unhindered to

Figure 9.3. Microstructure of a cast pure metal (magnification ×95)

give a large grain size. With a fast cooling rate many more nuclei are produced. This is because not all nuclei form at the same time. If the temperature drops rapidly, even to below the usual melting point, additional nuclei form before the original ones have had any real chance to grow. As a result, the growing nuclei soon bump into their neighbours and the final grain size is relatively small.

In contrast to pure metals, alloys usually solidify over a range of temperatures. Broadly speaking, the process of nucleation and growth is

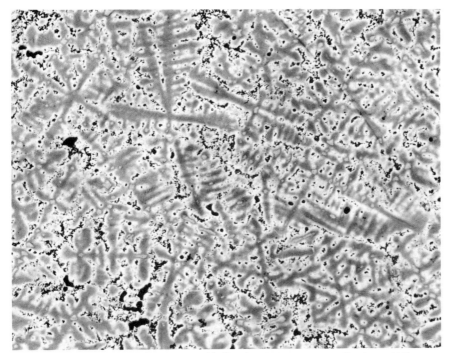

Figure 9.4. Microstructure of a cast 9 ct gold alloy showing coring (magnification ×200)

similar but metallographic examination of the cast structure may show the dendritic nature of solidification due to alloy compositional variations across the dendrite arms. This is known as 'coring' (Figure 9.4). Furthermore, more than one crystal phase may be present in the microstructure.

Solidification in a mould

When a molten metal or alloy is poured from a crucible into a mould, i.e. the process of casting, it begins to solidify from the mould walls (Figure 9.5). If a cold metal mould is used, as is the case in ingot casting, the rate of heat removal is rapid and initially a layer of tiny chill crystals is formed on the mould wall. Long finger-like grains (columnar grains) then begin to grow from the walls towards the centre of the ingot (Figure 9.6a). The extent of the

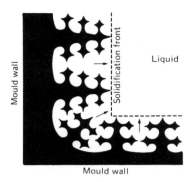

Figure 9.5. Nucleation and growth from a mould wall

columnar growth is dependent on the temperature of the melt. A low pouring temperature will allow nucleation to take place in the centre before the columnar grains will have had time to grow into this region (Figure 9.6b). When a plaster or sand mould is used, the cooling rate is slower and equiaxed grains are obtained throughout the casting (Figure 9.7).

The grain structure, grain size and ultimately the success of the casting depends very much on the pouring temperature and the mould temperature. If the pouring temperature is too low, the melt begins to solidify before the mould has been properly filled resulting in imperfect castings. On the other

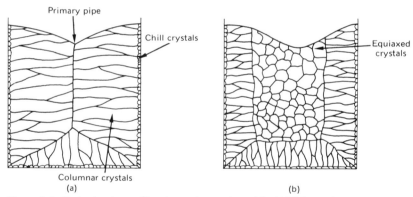

Figure 9.6. Grain structures of ingots cast in metal moulds. (a) High pouring temperature. (b) Low pouring temperature

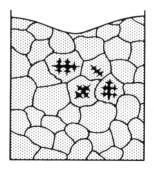

Figure 9.7. Grain structure of investment or sand casting

hand, if the pouring temperature is too high, excessive oxidation and absorption of gases from the surrounding air or furnace atmosphere can occur and, in addition, a coarse grain size is obtained. These effects may give trouble in any subsequent working operations or give brittle castings.

Recommended pouring temperatures are usually of the order of 75 to 100°C above the liquidus temperature.

Shrinkage

Metals and alloys undergo considerable shrinkage when they solidify (Figure 9.8). Gold, for example, contracts 5.1 per cent by volume on solidification. The practical consequence may be seen in Figure 9.6. As the melt solidifies

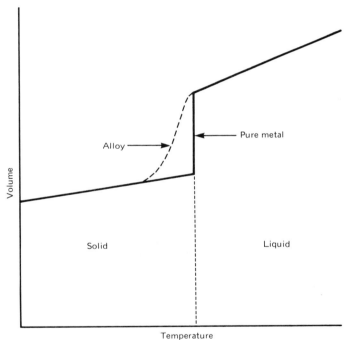

Figure 9.8. Contraction during solidification

from the mould wall towards the centre, a funnel-like depression known as a 'primary pipe' is produced at the top. With ingots which are going to be rolled or drawn, the pipe must be cropped off prior to working, otherwise a centre-line defect will be obtained in the wrought product. When producing cast shapes, sufficient provision must be made for a reservoir of molten metal so that shrinkage voids are confined to this region and liquid metal feeding into the casting during solidification can counteract contraction effects.

In a similar way, molten metal is trapped between dendrite arms as they grow during solidification, and it is important to be able to feed molten metal back into these areas during the growth process to prevent the formation of interdendritic porosity or shrinkage porosity. This becomes increasingly difficult with alloys that have a wide solidification range and also with certain shapes of casting. The presence of this type of porosity considerably reduces the strength of the casting.

Melting and alloying practice

Alloying is done at the melting stage. It is essential that the constituent metals are pure. Small traces of impurities such as lead, arsenic and antimony form brittle intermetallic compounds in the grain boundaries of cast carat golds which will cause catastrophic cracking on subsequent working. Similarly, the presence of sulphur causes problems with the nickel white golds due to the formation of a low melting point nickel sulphide (NiS).

Fortunately, gold, silver, copper, zinc and platinum are produced in a very pure state so that impurities are most likely to come from remelted scrap or from equipment such as crucibles and stirrers. Obviously a lot of scrap is remelted in the trade and it is important to ensure that only clean scrap is used. Metals have a characteristic relationship between their vapour pressure and temperature. Vapour pressure is a measure of the tendency for atoms or molecules to escape from a substance whether it is a liquid or a solid. A molten metal will gradually evaporate even below its boiling point in much the same way as a pool of water will slowly disappear after a rainstorm. The rate of evaporation increases with increasing temperature. Because of this, it is good practice to make additions of the higher melting point constituents to the melt of the lower melting point constituents during alloying if possible. Zinc is a special problem because of its relatively low melting point and high vapour pressure, and for those carat golds and silver solders containing zinc it is usually added as a pre-alloyed copper-zinc 'master alloy'.

The pure metals are often produced in granular form for alloy manufacture for ease of weighting out and melting.

Melting equipment

The types of furnaces and melting equipment are many and varied, ranging from a single gas-air torch assembly to gas- or oil-fired furnaces or to modern induction melting plant. Figure 9.9 shows typical layouts for the latter two methods.

In induction melting, the crucible and its contents act as the secondary component in a transformer. The primary water-cooled copper coil surrounds the crucible and the induced current in the secondary causes heating and melting. It is an extremely rapid form of heating and melting compared with other methods.

The main precaution to be taken is that of minimizing gas pick-up by the molten metal. This may come from the surrounding atmosphere or from moisture on the surface of tongs, crucibles and stirrers. One advantage of induction melting is that homogeneity in the melt is improved by electromagnetic stirring. Vacuum melting, using either electric resistance or induction heating, is becoming popular because of the complete absence of a surrounding atmosphere and the fact that the melt may even be degassed, giving sounder castings.

The requirements for crucibles are that (a) they should not react with their contents, and (b) they should have a high resistance to softening or cracking with sudden changes in temperature. Plumbago crucibles prepared from

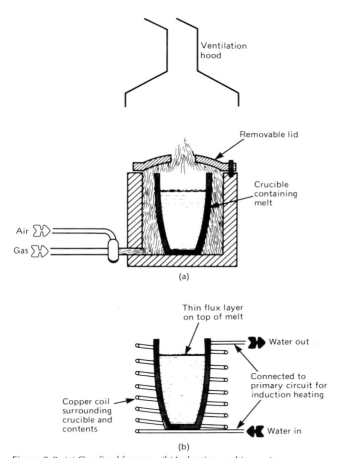

Figure 9.9. (a) Gas-fired furnace. (b) Induction melting unit

graphite and clay or pure graphite crucibles are recommended for silver and the coloured carat golds but fireclay crucibles are best for melting white golds since nickel may react with the carbon graphite in the plumbago, and in the higher melting palladium-containing alloys, silicon or sulphur pick-up may occur in the presence of carbon. The higher melting point platinum metals require aluminium oxide crucibles for robustness. Molten platinum alloys should never be melted in crucibles containing carbon.

Plumbago stirrers, if necessary, are best for the coloured carat golds, and iron rods may be used for silver. On no account should iron be used for gold because it will be dissolved in the melt.

Fluxes

A flux is frequently added during melting to (a) provide a protective cover on top of the melt and prevent oxidation and gas pick-up from the atmosphere, and (b) react with any base metal oxides in the melt to form a molten slag which floats on top of the melt. Only a minimum amount of flux should be used as there is a danger it may be trapped in the cast product during pouring as a glassy inclusion.

The fluxes commonly used for silver and gold alloys are borax which melts at 760°C and boric acid (melting point 870°C). The latter is preferred for the higher melting-range alloys, e.g. the palladium white golds.

Lumps of charcoal are often floated on the melt where they slowly react with the surrounding air to form a layer of carbon monoxide reducing gas giving further protection from oxidation. However, this technique should not be used with white golds or platinum alloys.

Other additives

Other additions are sometimes made to large melts. Refining fluxes such as ammonium chloride react with base-metal impurities to give chlorides which volatilize off into the surrounding atmosphere. Degassing tablets, e.g. hexachloroethane, volatilize and in doing so remove dissolved gases from the melt. Deoxidizers react with oxygen in the melt to form volatile oxides. They are particularly useful for silver which can dissolve a large amount of oxygen in the molten state. Cadmium (0.5 per cent) was formerly the recommended deoxidizer for silver but this is not advisable because of the evolution of poisonous brown fumes of cadmium oxide. Nowadays, small additions of phosphor-copper are used where the phosphorus reacts with the oxygen to give a volatile oxide. Care must be taken to avoid using excessive amounts as the presence of residual copper phosphide in the grain boundaries causes embrittlement.

Gold itself dissolves virtually no oxygen when molten, but this is not true of the carat golds because of their silver and copper content. Modern low-carat alloys usually contain zinc and sometimes also silicon, both of which act effectively as deoxidizers.

Casting processes

Here we are concerned with either the production of cast ingots for further fabrication into wire, rod, sheet, tube, etc., by working processes or the production of final shapes, i.e. castings, apart perhaps from a surface finishing treatment.

Ingot casting

The metal or alloy is poured into an ingot mould to produce a rod, bar or slab depending on the eventual wrought product. Ingot moulds are generally cast iron, steel or water-cooled copper moulds in two halves which are clamped together during pouring but which can then be split for easy removal of the solidified ingot.

Vertical moulds are preferred with the molten metal stream poured centrally in through the open top. It is important to stop excess flux, charcoal lumps and oxide dross from entering the mould, otherwise they may be trapped giving a defective ingot. To prevent sticking to the mould walls, a mould dressing is applied before pouring, e.g. a layer of lampblack, tallow, beeswax or machine oil. Moulds should be preheated to prevent gas pick-up from any moisture that may be present.

Because of the nature of the precious-metal industry and the high material cost, ingots are relatively small compared with those produced normally in the metallurgical industry.

It is general practice to strip the moulds and quench carat golds and silver-alloy ingots very soon after they have solidified, and while they are still hot, in a pickle to brighten the surface and to remove any flux or charcoal imbedded in the surface. Quenching is also necessary for the carat golds to obtain a soft ductile grain structure for good workability and to avoid hardening by precipitation and ordering reactions which would otherwise occur on slow cooling.

More recently, some of the larger companies have used the continuous casting process for large-scale production of gold alloy and sterling silver bar. Here, molten metal is poured from a holding furnace into either a water-cooled vertical or horizontal mould while the solidifying bar is withdrawn continuously by extractors rolls from the other end of the mould where it may be cut into suitable lengths for subsequent working.

Production of castings

There are a wide variety of methods for producing cast shapes but we shall only look briefly at the two most commonly found in the precious metal industry.

Lost-wax investment casting

The lost-wax process, often referred to as 'cire perdue', is now widely used in the jewellery trade and yet it has had rather a chequered history. It has certainly been in use for at least 4500 years. There are beautiful examples of lost-wax castings made by such varied cultures as the Siberian nomads of Kazakhstan, the Sumerians at Ur in the Middle East, the Greeks, the Egyptians and the Chinese. At a later date, starting about 400 AD, the process was developed independently in the New World by the pre-Columbian Indians of South America and the Aztec and Mayan civilizations in Central America. One of the most notable exponents of the art was the famous Italian goldsmith Benvenuto Cellini (16th century) who made a number of improvements to the technique and left written records of his methods. Then for some inexplicable reason the process largely fell into disuse for about 400 years with the exception of its use by Carl Fabergé who made exquisite jewellery for the Russian court around the turn of this century. Surprisingly, the dental profession was mainly responsible for the re-emergence of the process as it was discovered about 1910 to be an ideal method for making gold teeth and crowns. During the 1930s it was taken up again by jewellers and today it is the single most important production process for jewellery manufacture. A number of major refinements have been introduced, the most notable being the incorporation of centrifugal casting and vacuum-assisted casting.

To describe the process, let us consider the mass production of gold ring castings.

1. A master pattern of a ring is made from a soft metal which can be relatively easily carved. Fine silver is ideal for this purpose. Alternatively, a wax model can be carved and used to make a casting for the metal master.

2. The metal master complete with a casting sprue is encased in a raw rubber pack which is vulcanized by heating in a press at about 150°C.

3. The solid rubber mould is carefully slit into two halves using a surgeon's scalpel and the metal master extracted.

4. The split halves are clamped together and hot wax is injected into the mould and allowed to cool. The solid wax reproduction of the ring is removed from the mould. Any number of wax reproductions can be made, depending on the production run. Each wax ring has the casting sprue attached usually to the ring shank opposite the section containing the gemstone mounting. The sprue is the channel through which the molten metal will run during casting.

5. The tip of each sprue is welded on to a larger central wax column (Figure 9.10) to produce a 'tree'.

6. An open-ended cylinder is placed around the wax tree and a slurry of investment compound is poured in over the tree. Modern investment powders, based on 'plaster of Paris' or phosphate-bonded silica (for higher casting temperatures), are highly sophisticated and have special additives and modifiers. Their composition is dependent on the temperature of the molten metal, i.e. relatively low temperatures for coloured carat golds and sterling silver, or high temperatures for palladium white golds and platinum alloys.

7. Before setting, the cylinder and its contents are often placed in a vacuum chamber to remove any air bubbles that may be trapped in the investment.

8. The hardened investment is placed upside down in a dewaxing furnace so that the wax melts and runs out at about 200°C leaving a hollow mould. The temperature is raised to 700–950°C, depending on the type, to fire and completely harden the investment. Finally, the temperature is brought to that required for casting, which may be from 350–800°C depending on the section thickness of the casting and the alloy being cast.

   500–600°C for 9 ct coloured gold
   600–700°C for 14, 18 ct coloured golds and silver
   700–800°C for white golds, palladium and platinum.

9. If centrifugal casting is to be employed, the metal is melted in a special crucible either by torch melting or by induction melting, the hot mould is placed behind the crucible on one arm of the centrifuge and the whole assembly is then spun around the central axis at high speed. Due to the centrifugal action the molten metal climbs up the crucible wall and through a hole into the mould (Figure 9.11).

Figure 9.10. Photograph of a tree with wax ring patterns connected by the casting sprues to a central wax column (courtesy Chris Walton, Worshipful Company of Goldsmiths)

10. With vacuum-assisted casting, the metal is poured from the crucible into the mould which is being subjected to a vacuum from below so that the melt is forced into all parts of the mould cavity by atmospheric pressure.
11. Finally, the investment compound is removed from the solidified casting and each ring is removed from the tree, the sprues cropped off from the ring shanks and final cleaning and polishing done.

The main advantages of the process are that reproduction of fine detail is excellent, complex shapes with undercuts and thin sections can be made

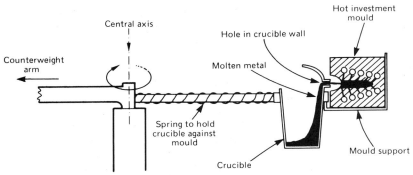

Figure 9.11. Centrifugal casting machine

which would be impossible by any other casting process, and large production runs are possible. Alternatively, the process is also ideal for one-off jobs where a carved wax model can be invested directly.

The central column which has acted as the reservoir to counteract shrinkage during solidification and the casting sprues are obviously precious-metal scrap which can be returned for remelting.

Sand casting

In this process, a master pattern of wood, metal or any material which can be shaped is packed around with a suitable moulding sand. The sand mould is made in two halves within metal surrounds so that they can be separated to remove the pattern prior to casting. A running and gating system through which the molten metal runs into the hollow cavity is incorporated in the sand mould. Venting passages to allow air to escape during pouring may also be cut into the sand. A number of small objects can be cast simultaneously in one mould by attaching all the patterns to a central runner with individual gates (sprues).

The process is not used for jewellery today, and therefore is not generally used for gold or for the platinum metals. However, it is used in the silverware trade for large cast objects and knobs, feet, spouts and handles of coffee and teapots both in sterling silver and in nickel silver for subsequent plating.

Casting defects

The production of successful castings depends on a large number of factors. The correct investment material or moulding sand must be used and the recommended pouring temperature for the alloy being cast adhered to. This latter factor will either depend on the skill of the caster in judging the right time to cast or it will necessitate the use of a temperature-measuring device, e.g. a dip thermocouple or optical pyrometer, to record the temperature of the melt. Too low a temperature and the mould will be incompletely filled.

The design of the casting and that of the sprueing or running and gating system must be considered if shrinkage cavities and gross interdendritic porosity are to be avoided. This means that the mould must be completely filled and solidified before the metal in the sprue solidifies and cuts off liquid-metal feeding.

Another defect which is due to tensile stresses set up by mould restraint as alloys contract on solidification is a 'hot tear', and it shows up as a crack along grain boundaries where the last traces of molten metal exist. It is more likely to occur in castings with a large grain size. This is quite common in sterling silver. In this case the remedy is to add a very small piece of high-carbon steel, e.g. the tip of a file, to the melt when the dispersed iron carbide particles act as nucleating agents to give more nuclei and hence a smaller grain size. Similarly, it is known that small additions of iridium, rhodium or cobalt silicide to carat golds will act as grain refiners, although this is not usually done in practice.

As indicated earlier, great care must be exercised in preventing or minimizing the dissolution of gases such as oxygen, hydrogen and water vapour in the melt and the mechanical entrapment of air during pouring. If this occurs, either the gas comes out of solution on solidification and forms trapped spherical 'gas blowholes' giving a weak casting or it reacts with alloy constituents to form oxide particles. This can be a problem with sterling silver since molten silver can absorb large quantities of oxygen and water vapour and these can then react with the copper content to form copper oxide inclusions. Rapid heating and melting by induction melting, vacuum melting or the use of a suitable flux help greatly in this respect.

Occasionally castings are found having poor surfaces or oxide inclusions at or just below the surface. This is due either to a reaction between the metal and the mould material or a reaction with the mould dressing or because the mould wall has crumbled as the metal rushes into the cavity.

One interesting defect which can occur in centrifugal casting is the presence of smooth hemispherical pores on the surface of one side of the casting in relation to its position on the tree. It has been shown that this type of defect arises from the use of excessive speeds in the casting machine and is caused by turbulence in the molten-metal stream.

There are other defects which are often specific to a particular process. Expert advice should be sought in such cases as the remedy may be simple.

10
Working and annealing practice

The main objective of a working process is obviously to produce a particular shape, be it a semi-fabricated shape such as rod, sheet and wire, or a more intricate finished article fashioned by a goldsmith.

Working can be done either at temperatures where annealing processes occur almost simultaneously, i.e. hot-working, or by cold-working at relatively low temperatures, usually room temperature, in which case a separate annealing treatment is necessary. Annealing is the process by which softness and ductility may be restored and by which a new completely strain-free microstructure is obtained.

A second objective of working is to destroy the original cast microstructure and replace it with this strain-free structure because in general metals and alloys have better strength, ductility and toughness in the wrought state than in their cast state. The reasons for this are as follows.

1. The grain size is usually smaller and more uniform after hot-working or cold-working and annealing,
2. Any shrinkage porosity tends to be closed up during working,
3. There will be a better and more uniform distribution of phases in multiphase alloys, and
4. Any traces of impurities are likely to have a less deleterious effect on mechanical properties.

To give an example, it is not uncommon for the claws in investment cast ring mounts to break during the setting of gemstones whereas ring mounts fabricated from wrought alloys are stronger and less liable to breakage.

Large ingots are initially hot-worked because this reduces the power and load requirements of the processing equipment. The final stages of fabrication are usually done by cold-working as it is easier to control dimensions and surface finish.

Because ingots of carat golds and silver alloys are relatively small and they are ductile it is common practice to directly cold-work after casting. Ingots of platinum, palladium and rhodium are usually hot-worked at about 1000°C prior to cold-working operations. As stated in Chapter 8, iridium cannot be cold-worked.

Work-hardening

When metals and alloys in either the cast or wrought and annealed conditions are cold-worked, they become stronger and harder and this is accompanied by a gradual loss in ductility (Figure 10.1). This phenomenon is known as work-hardening or strain-hardening. If cold-working is

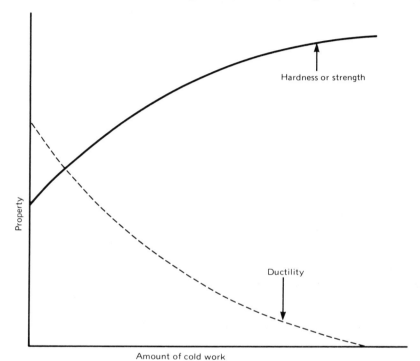

Figure 10.1. Effect of cold-work on mechanical properties

continued, fracture will eventually occur because the metal will completely lose its ductility.

During cold-working, the original grain structure of the metal becomes distorted, the grains gradually elongating along the direction of metal flow. At high levels of cold-work, the grain boundaries lose their identity as the grains become heavily distorted and a typical fibrous structure is obtained (Figure 10.2).

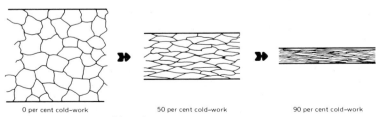

0 per cent cold-work   50 per cent cold-work   90 per cent cold-work

Figure 10.2. Effect of cold-work on microstructure

A major cause of fracture during fabrication is overworking. Other causes are 'hot shortness' and 'cold shortness' due to the formation of low melting point or embrittling phases.

Obviously, during the manufacture of wrought products, whether the semi-fabricated products or the more intricately-shaped final products of the gold- and silversmith, it is essential to avoid overworking the metal and to restore its softness and ductility by annealing. The amount of cold-working permissible before annealing becomes necessary depends on the work-hardening characteristics of the metal or alloy in question.

With ductile metals like the carat golds, silver alloys, platinum and palladium it is good practice to anneal after about 70 per cent strain as measured by percentage reduction in thickness or area on working. However, this advice can be followed strictly only in the manufacture of semi-fabricated products. When a craftsman bends or shapes an article of jewellery or decorative metalware the amount of strain introduced cannot easily be quantified. Experience must then decide when annealing is necessary. Rhodium is less ductile and can be only moderately cold-worked before annealing is required.

Annealing

The effects of increasing annealing temperature on the properties of a metal that has been cold-worked are summarized in Figure 10.3, and the accompanying changes in microstructure are illustrated below the graph. As the temperature is raised, at first certain recovery processes take place. The

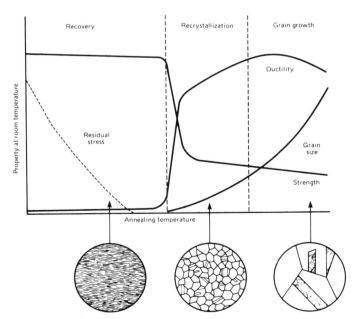

Figure 10.3. Effect of annealing temperature on properties and microstructure

high strength and low ductility resulting from cold-working are virtually unaltered. There is little change in the microstructure, and annealing does not occur. However, a reduction in the level of residual or internal stress can be achieved in this temperature range, i.e. a treatment known as stress-relief annealing. The subject of residual stresses which may be introduced into a metal by non-uniform working or by non-uniform heating and cooling, and which can have important consequences for some of the carat golds, will be discussed later in this chapter.

The important temperature range for full annealing is the recrystallization range (Figure 10.3). It can be seen that during full annealing the strength decreases to a much lower level and the ductility rises to a high value. The cold-worked fibrous structure is replaced by a new strain-free equiaxed grain structure having a relatively small grain size.

As the temperature is raised yet higher, the strength continues to decrease slightly and the ductility increases to a peak value after which it may fall as the solidus temperature is approached. At this stage, the most noticeable change is the rapid increase in grain size. It is generally advisable to avoid annealing in this grain growth range because the presence of large grains in the metal can lead to marked surface unevenness on subsequent working. This is the 'orange peel' effect, so-called because of the similarity of the metal surface to that of orange peel.

The factors which affect recrystallization and annealing behaviour are:

1. The amount of prior cold work

The greater the amount of cold work, the smaller will be the resulting grain size on annealing and the lower the required recrystallization temperature.

2. Temperature and time of annealing

Longer times are required at lower annealing temperatures within the recrystallization range but temperature changes have a far greater effect on the rate of recrystallization. In practice, if annealing is done in a furnace, a temperature is selected where annealing is complete within say 30 minutes. With torch annealing, where times are obviously much shorter, slightly higher temperatures are required.

Composition

The recrystallization temperature is strongly dependent on metal or alloy composition. In the case of pure metals, a good guide is to assume a temperature which is approximately one third of the melting temperature expressed in degrees absolute (Kelvin), i.e. melting point °C + 273 = melting point K. Pure lead will recrystallize at room temperature and can only be cold-worked at sub-zero temperatures whereas pure gold recrystallizes at 200°C. The presence of impurities can raise the recrystallization temperature by an appreciable amount, and it follows that alloying will have a similar effect.

Typical annealing temperatures for the precious metals and alloys are given in Table 10.1.

Control of grain size and mechanical properties in annealed material is best achieved by the use of furnaces and accurate temperature measurement. However, for jewellers and silversmiths this is rarely convenient, and torch-annealing with a gas-air flame is and will continue to be widely used. Articles to be torch-annealed are placed on a charcoal block or on a loose bundle of iron wire flattened to a circular plate for heating.

Judgement of correct annealing temperature is made by assessing colour changes in the heated metal by eye. A list of colour temperatures is included in Table 10.1. It is important to do torch-annealing in a darkened corner of the workshop if consistency is to be attained, for one's judgement of colour temperature is seriously impaired by reflections of light from windows, fluorescent lighting, etc.

The reducing part of the flame should be used to keep surface oxidation of the base metal content in carat golds and sterling silver to a minimum. Items should be allowed to cool to black heat, i.e. when no reddish colour due to heat is observed, quenched into cold water and then pickled in hot

Table 10.1. Typical annealing temperatures

Material	Annealing temperature	Colour
Fine gold	200	Black heat
Fine silver	200	Black heat
22 and 18 ct coloured golds	550–600	Very dark red
14 and 9 ct coloured golds	650	Dark red
Pd white golds	650–700	Cherry red
Ni white golds	700–750	Cherry red
Sterling silver	600–650	Dark red
Platinum	800	Bright red

sulphuric acid. The object of this is twofold. First, any oxide scale will be removed by pickling. Secondly, further changes in the microstructure due to ageing and ordering processes (see Chapters 6 and 7) can occur on slow cooling which will harden the alloys, and this will be undesirable if further working is to be done. Rapid quenching prevents the ageing and ordering reactions and retains the maximum softness and ductility imparted by annealing.

Nickel white golds should be air-cooled on an iron plate after annealing and then cold-pickled as rapid quenching introduces high residual stresses. These are undesirable as they can lead to subsequent cracking in the article.

Pickling solutions are usually acid-based, e.g. 10 per cent sulphuric acid in water, which dissolve the base-metal oxides from the surface. Since nickel oxide does not readily dissolve in dilute acid pickle, special pickles containing sodium or potassium dichromate are necessary.

When furnace annealing is employed, a protective non-oxidizing atmosphere can be supplied, e.g. cracked ammonia or nitrogen—10 per cent hydrogen mixture known as 'forming gas', which prevents oxide scale formation thereby minimizing the need for pickling. This process is referred to as 'bright annealing'. Quenching may still be necessary to prevent ageing and ordering.

An alternative is salt-bath annealing in which articles to be annealed are placed in a basket and lowered into a hot molten salt mixture contained in an iron pot. The benefits are that the equipment is cheap, temperature can be easily controlled, heating of the articles is rapid and they are protected from oxidation. Salt baths are particularly advantageous, though little used, in the manufacture of small pieces of sterling silver such as tableware.

Working processes

The main processes used for producing semi-fabricated products, i.e. products which will be finally shaped by further working, are sheet and rod rolling, wire drawing and tube drawing.

With carat gold and sterling silver sheet production, it is general practice to directly cold-roll the stripped slab ingots using a powerful electrically driven mill consisting of a pair of forged steel rolls (Figure 10.4). A number of rolling passes are needed to obtain sheet or strip of the required thickness and inter-stage annealing may be necessary before this is achieved. In commercial production the sheet or strip will be coiled and the coils annealed in a protective atmosphere furnace.

For rods and bars, square-section ingots are rolled using a pair of rolls having a series of grooves of decreasing dimensions. As mentioned, platinum and palladium ingots are usually hot-rolled prior to cold rolling and annealing.

In wire production, rods are rolled down to about 3 mm diameter and further reduced by drawing through a series of carbide dies (Figure 10.4). Wax, tallow or soap is used as a drawing lubricant. Small workshops which require short lengths often use hardened steel draw plates containing die orifices in a wide variety of shapes and sizes.

Although there are a number of techniques for producing seamless tubing, it is common practice in the precious-metal trade where only short lengths are wanted, to deep-draw a cup from a circular piece of sheet and then to redraw the cup to increase its length and reduce the wall thickness (Figure 10.4). Finally, after a number of redrawing stages, the tube is transferred to a wire drawing bench for reduction to small-diameter tubing.

The manufacture of finished articles will incorporate the use of one of more of these semi-fabricated products and, of course, may also include castings, e.g. handles, knobs, feet, etc.

In considering the manufacture of sterling silverware and nickel silver for subsequent plating, we can conveniently distinguish between flatware and holloware. With flatware the principal working process is forging. Taking spoons as an example, blanks are stamped out from thick sheet. After some preliminary hammering and rolling operations to impart initial shaping, handle ends are placed between hardened steel dies shaped to the required design and formed using a drop hammer or a hydraulic press. The excess metal (the 'flash') exudes from the sides of the die and is trimmed off. Spoon bowls are produced by hammering the other end of the blank using a shaped punch and die assembly.

86

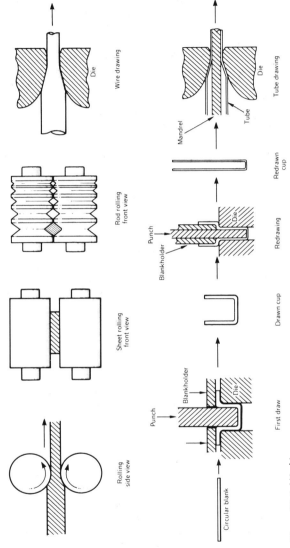

Figure 10.4. Working processes

Holloware is essentially produced from sheet metal and includes tea and coffee pots, goblets, bowls, sauce and gravy boats, etc. Traditionally, these have been made by hand raising in which a circular piece of sheet metal having a diameter approximately equal to the width plus the depth of the desired article is hammered at right angles to the surface over a series of shaped stakes. At each stage of the raising operation the metal work hardens and periodically needs to be annealed. The hammer marks are removed by planishing with a special type of hammer leaving a relatively smooth surface which is then finally polished.

Round items, e.g. goblets and bowls, can be made by the process of metal spinning. Here, a circular piece of sheet metal is clamped between a chuck and a follower and the whole assembly rotated in a machine similar to a lathe. Pressure is applied to the back of the spinning sheet by a steel tool so that it is gradually bent over and around the chuck which is shaped to that of the desired article. Chucks are fashioned from wood, steel or nylon.

Holloware may be made also by stamping or by deep-drawing and stretch-forming in a shaped punch and die assembly using either a drop-hammer forge or a press.

Decorations such as mount wires, castings, etc., are then soldered to the shaped item. However, much of the beauty of holloware is imparted by surface decoration. These techniques include embossing, flat chasing, repoussé chasing and engraving, and highly skilled people are required for this work.

Gold and platinum ware, apart from jewellery, can be made by methods similar to those used for silverware but naturally such items are not produced in the same quantity because of the cost of metal involved. One notable exception to this is the use of hot stamping (forging) in the manufacture of carat-gold watch cases.

Some specific problems

Before leaving the subject of working and annealing practice, it is pertinent to look at some specific problems that can arise during fabrication.

'Firestain' in sterling silver

Reference has already been made to the formation of base metal oxides on the surface of precious-metal alloys during annealing in an oxidizing atmosphere such as air. This can be a particular problem with sterling silverware. Oxygen not only reacts with copper to form oxide scale on the surface but it diffuses into the alloy and forms tiny particles of copper oxide below the surface. The result is that, although pickling will remove the surface scale giving a very thin layer of essentially fine silver, subsequent polishing reveals a dark mark as the sub-surface oxide particles become exposed. The higher the annealing temperature or the longer the annealing time, the greater is the depth of penetration of oxygen and the more persistent is the firestain effect.

When it occurs, it can only be removed by electrolytically stripping-off the surface layers or by heavy abrasion and polishing. Alternatively, it may be covered by a thin electroplated coating of silver or rhodium.

However, borrowing the medical adage that 'prevention is better than cure', the problem may be avoided by salt-bath annealing or by bright annealing in a furnace with a protective atmosphere. Where torch annealing is used, articles should be covered with pieces of charcoal or a protective flux layer and excessively high temperatures and long times avoided. Alternatively, a gas-flux dispenser may be incorporated in the gas line of the torch.

The incidence of firestain is not confined to annealing since it can occur during soldering operations and also on the surface of castings.

'Fire-cracking' in nickel white golds

When metals are cold-worked, the deformation is not distributed uniformly throughout the material, i.e. the amount of strain varies from point to point within the article. The result is that the article becomes internally stressed. The magnitude of these internal or residual stresses depends on the differences in localized strains and on the mechanical properties of the metal. The stresses may be tensile or compressive depending on the type of deformation received.

Usually, residual stresses introduced during working are relieved in early stages of annealing and are not a problem. With the nickel white golds, particularly ring shanks, the stress level can be very high after cold work and if they are rapidly heated to full annealing temperatures of about 750°C the tensile internal stresses may be sufficient to cause crack formation in the article. 'Fire-cracking', as it is called, may be eliminated by slow heating to about 300°C, to allow stress relief while the alloy is still strong enough to resist cracking, and then heating up to 750°C.

The problem can also occur in the copper–nickel–zinc alloys (the nickel silvers) and so it is probable that the presence of both nickel and zinc is partly responsible as zinc is an alloying constituent in the nickel white golds.

Residual stresses can also be introduced by uneven heating and cooling, particularly cooling, due to differences in the thermal expansion and contraction within the article. The stress level increases with increasing rate of cooling and this is why the nickel white golds should not be quenched after annealing and soldering.

Stress corrosion cracking in carat golds

The problem of stress corrosion cracking in certain 9- and 14-carat gold jewellery is often baffling to jewellers because it may occur not only during fabrication but also during subsequent storage or even after sale to a customer (Figures 10.5 and 10.6). In fact, the basic causes of this embarrassing phenomenon are relatively simple to appreciate.

As the name suggests, cracking only occurs when stress and corrosion are combined, either by subjecting the alloy to a stress while it is in a corrosive

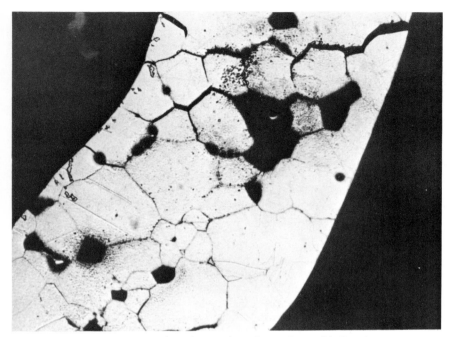

Figure 10.5. Stress corrosion cracks in the grain boundaries of 9 ct gold-alloy chain (magnification ×300)

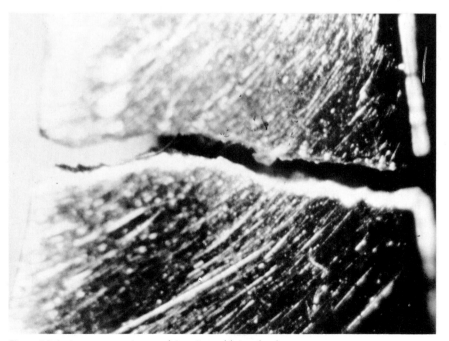

Figure 10.6. A stress corrosion crack in a 9 ct gold ring shank

environment or, as is often the case, by exposing it to a corrosive environment when it is already in an internally-stressed condition. Yet in the absence of stress the article would not suffer from corrosion.

Scientific investigation has shown that it only occurs in alloys where the gold content is less than 40 atomic per cent, i.e. less than 40 per cent of the total number of atoms in the alloy are gold. Hence, 18- and 22-carat golds are immune from this form of attack. The corrosive environment is supplied by reagents such as acids and chlorides. The fact that these reagents are common constituents of pickling baths gives a major clue to the reasons behind this type of failure during fabrication, the residual stresses having been introduced by prior working or quenching operations.

Cracking in storage can often be traced to fumes from nearby pickling tanks. Industrial atmospheres contaminated by dust, ammonia or chloride fumes can explain the occurrence after sale to a customer. Chlorine in the water in swimming baths has been known to cause stress corrosion cracking in 9-carat gold.

The potential risk of stress corrosion cracking on finished jewellery may be avoided by a full anneal followed by slow cooling to room temperature. If a final work-hardened condition is preferred for added strength, the alternative is to apply a brief stress-relieving treatment at 250–300°C. As a result, the internal or residual stresses are eliminated while the work-hardened state is essentially preserved.

Polishing and finishing treatments

The importance of polishing and finishing as part of the fabrication process cannot be overemphasized because of its contribution as a selling factor.

The polishing and finishing treatments for silverware can broadly be separated into three stages. The first is a fine grinding process and it involves the removal of metal from the surface of the article by abrasion. Silver articles are sanded to remove surface marks left by working such as in planishing, stamping and spinning. This is done with sand (or pumice) in oil using a revolving wheel of felt, leather or walrus hide. The finish is influenced by the grade of sand and pumice used.

The second stage is buffing in which the surface layer actually spreads out or flows so that the minute scratches left by grinding are gradually filled up. The polisher uses mops of calico or linen and cotton and a compound such as 'tripoli'.

The third stage is finishing using softer mops with jeweller's rouge mixed with water, methylated spirit or paraffin.

The process requires a high degree of skill, and a wide variety of finishes can be produced depending on the grinding and polishing techniques used. Collectors of silverware will be aware that the lustre is influenced by the technique adopted. Many manufacturers aim at a 'black' finish which is described as 'antique'. Another popular finish is 'butler finish', so-called because it resembles that produced by butlers polishing the family silver using a heel of hand application of rouge and whiting mixed with water.

Sanding is not common for goldware for the simple reason that too much valuable metal would be removed by the abrasive process. Consequently, large pieces are only buffed and finished with tripoli and rouge.

It will be appreciated that hand polishing of items of jewellery mass-produced in large numbers would be a very tedious business, e.g. rings, charms, chain, etc. Nowadays, the grinding, often referred to as cutting, and polishing are done by barrel polishing or vibratory polishing.

In conventional barrel polishing, the parts are tumbled in rotating barrels. Three operations are involved: cutting, colouring and burnishing. In cutting, the parts are loaded into the barrels with plastic cones, water and a cutting compound. After tumbling for 8–12 hours the barrel and contents are thoroughly rinsed with water. At this stage the parts are smooth but dull. The purpose of the colouring operation is to brighten the parts prior to burnishing using plastic cones and a proprietary compound. After 1–2 hours the parts are again thoroughly rinsed. In burnishing, which removes no metal as distinct from the previous two operations, tumbling is done using steel shot and a burnishing compound for 24 hours. After a final rinse the parts are dried in a heated tumbler using corn-cob as the drying medium.

More recently, centrifugal tumbling has been introduced in which centrifugal forces are generated by rotating the barrel around the external axis. The advantages are that processing times have been reduced to 1–2 hours, there is less loss of gold and the surface finish is superior to that of conventional tumbling.

In the USA, the use of vibratory bowl-type polishers has gained wide acceptance. The cutting and burnishing operations are created by cascading and a horizontal movement that keeps the media and jewellery in contact for 100 per cent of the time and the action is said to duplicate that of hand polishing. Silicon carbide, emery and quartz preforms are used for deburring and coarse cutting and silica flour for fine cutting. Burnishing is done with steel shot. Treatment times are one-quarter to one-half that taken for conventional barrel polishing.

11
Joining techniques

It is commonly found that an article has to be made which is so complex in shape that it is either impossible or too costly to fabricate it directly in one piece. However, it is feasible to produce such an item by assembly of several cheaply-produced parts of simple design and joining them together by a suitable process. This applies right across the spectrum from industrial engineering components and structures to small intricate pieces of jewellery and silverware, e.g. ring shanks, filligree brooches, chain links, attachment of jewellers findings, incorporation of gemstone mounts, spouts and handles of coffee pots, etc. The main types of joining processes are grouped into the following:

mechanical methods – these include rivetting, bolting, screwing
welding
adhesive joining
soldering

Mechanical methods should be considered only when adhesive joints are not possible or would lack the necessary strength, and when welding or soldering may introduce distortion and internal stresses due to localized heating.

Soldering is by far the most common joining procedure used by the jewellery and silverware trade and the principles and practice will be discussed in some detail.

Principles of soldering

The process has been known from earliest times by goldsmiths. Both gold and silver soldering were practised in Mesopotamia before 3500 BC. The famous Leyden Papyrus found in Egypt and now in the Museum of Antiquities, Leyden, Holland, was written in Greek c350 AD. It outlines methods for making solders.

It is important to draw a distinction between soldering and brazing. The latter term is used in the engineering industry for soldering operations at relatively high temperatures, say above 600°C, using what are known as

92

brazing or hard solders which give strong joints, whereas soft solders are used at relatively low temperatures and are not particularly strong. The most commonly-used soft solders are based on lead-tin alloys, e.g. plumber's solder and radio or tinman's solder. The jewellery and silverware trade use the term soldering to apply to all temperatures and, in truth, the principles are the same in all cases.

These principles may be listed as follows:

1. There is no melting of the surfaces of the parent metal or alloy pieces being joined.
2. The solder metal or alloy has a lower melting point or range than that of the parent metal. In some cases the solder may have a completely different composition and this isn't a problem, e.g. silver solder for titanium. To conform to hallmarking regulations the solder must be similar and this presents certain difficulties. In practice there should be a difference of at least 50°C between the temperature at which the solder will flow and the solidus temperature of the parent metal.
3. Joining occurs by the action of capillary flow in the joint gap. In this connection it is important that the gap is small to ensure that capillary flow takes place throughout the length of the joint.
4. The molten solder must wet the parent metal surface, rather in the same way that water completely wets a clean grease-free surface, and does not form individual droplets having a high contact angle (Figure 11.1).

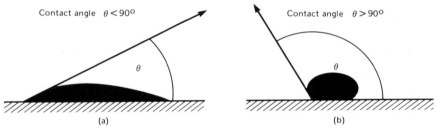

Figure 11.1. Molten solder should wet the metal surface, as in (a), and not form individual droplets having a contact angle, as in (b)

5. A good strong bond must be formed at the interface between the solder and the parent metal. This is usually achieved by a small amount of alloying across the interface by diffusion processes.
6. To ensure good wetting and bonding, it is essential to have a clean oxide-free and grease-free parent surface. This is obtained by the use of a suitable flux which dissolves the oxide film and dirt and allows the molten solder to contact the cleaned surface.

Joint design

The importance of a small joint gap has already been emphasized. In designing a joint to be soldered, it is advisable to arrange for the solder to flow from one end or side only so that it drives the air and molten flux ahead of it and completely fills the gap thus avoiding porosity and points of weakness due to entrapment of flux.

The solder joint must be mechanically strong, particularly as the solder itself may be weaker than the parent metal. Butt joints are the weakest but may be aesthetically pleasing for an item of jewellery. If a straight joint is required, a scarf joint should be made as it is stronger and yet looks similar to a butt joint in the finished article. Lapped joints are strongest but may not look attractive. Figure 11.2 shows examples of good and bad joints.

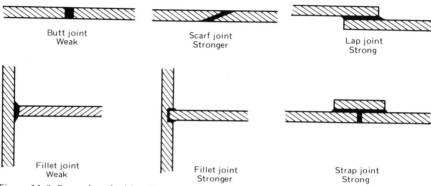

Figure 11.2. Examples of soldered joints

In the manufacture of jewellery and decorative metalware it is desirable that solder of a similar colour is used, otherwise the joint may look unsightly. In addition, it is a requirement of the Hallmarking Act that the solder must have a similar precious-metal content to that of the item being soldered wherever possible. When it is not possible to have a solder with a sufficiently low melting range and of the same precious metal content, then the hallmarking regulations stipulate what the minimum fineness of the solder should be.

Where an item of jewellery or silverware is made with a number of soldered joints, a stepwise approach is used. The first joint is made using a solder of high melting range, and then as subsequent joints are made the melting range of the solder is lowered to avoid remelting the joints made earlier in the process. The terms used in the industry to describe these melting ranges are 'hard' for the highest range, 'medium' for an intermediate range, and 'easy' and 'extra-easy' for the lowest ranges. Enamelling-grade solders with a high melting range are recommended for articles which will subsequently be enamelled again because of the necessity to avoid remelting the joint when the enamel is fired in a kiln.

Silver solders

A wide range of silver-solder compositions exist for industrial brazing applications such as for copper-base alloys, carbon steels, stainless steels, titanium, etc. These solders usually contain less than 50 per cent silver. One very popular series are the Easy-Flo solders.

For hallmarked silver, the Act states that silver solders containing not less than 65 per cent Ag must be used. The lowest melting temperature, i.e. the solidus, in the Ag-Cu system is 779°C and this is too close to the solidus temperature of sterling silver for simple Ag-Cu solders to be used without the risk of melting the parent metal. Additions of zinc lower the solidus and melting range, and for the extra-easy grades further additions of cadmium or tin may be made. Tin is now preferred to cadmium because of the danger of

Table 11.1. Silver solders

Grade	Silver %	Copper %	Zinc %	Melting range °C
Enamelling	71	22	7	730–800
Hard	80	13	7	745–778
Medium	70	22.5	7.5	720–765
Easy	65	22	13	705–723

cadmium oxide fumes. Table 11.1 shows typical silver-solder compositions and their melting ranges which are suitable for articles being hallmarked.

Unless skilfully made, the joint may be visible because the higher copper content gives a pronounced yellowish tinge to the solder.

Gold solders

With the coloured and white carat golds there is more flexibility in selecting a composition which will have the same caratage but a lower melting range than that of the piece being soldered. For example, reference to Figure 7.5

Table 11.2. Gold solders

Grade		Au %	Ag %	Cu %	Zn %	Cd %	Pd %	Ni %	Melting range °C
18 ct	Hard	75	12	8		5			826–887
	Hard	75		15	1.8	8.2			793–822
	Medium	75	9	6	10				730–783
	Easy	75				unknown			633–705
14 ct	Hard	58.5	25	12.5		4			788–840
	Hard	58.5	8.8	22.7		10			751–780
	Medium	58.5	4.9	25.6	2	9			738–760
	Easy	58.5				unknown			703–730
9 ct	Hard	37.5				unknown			756–793
	Medium	37.5				unknown			720–760
	Easy	37.5				unknown			695–715
	Extra-easy	37.5	20	20.5	1	18	3		630–690
White golds		80			8			12	782–871
14 ct		58.3	15	5.7	15			6	704–736
9 ct		37.5				unknown			710–743

Data from 'Gold Usage' by W. S. Rapson and T. Groenewald, Academic Press 1978, and from a catalogue issued by Johnson Matthey Metals Ltd.

shows that the liquidus temperatures vary widely for Au–Ag–Cu alloys of the same caratage and this is also true for the solidus temperatures. Therefore, it is often possible to select a self-type solder. The problem here is that although the caratage is the same, the colour may differ because of the need to select a composition with a lower melting range as comparison of Figures 7.5 and 7.6 will show. Subsequent hard gold plating will disguise colour differences in the finished article. To achieve the lower melting ranges for easy and extra-easy grades, zinc, cadmium, palladium and tin additions may be made. These additions will also modify the colour and may make it easier to match that of the parent carat gold. Manufacturers obviously prefer not to divulge exact compositions but Table 11.2 gives an indication of typical solders including some white gold solders.

Platinum solders

Solders recommended for platinum ware contain combinations of platinum, palladium, gold, silver and copper as well as other alloying additions. Normally white Pd–Ag–Au and Pt–Ag–Au alloys with melting ranges lying between 1200 and 1600°C are used but fine gold itself can be used if the colour difference is not important. The only stipulation for hallmarking platinum is that the precious metal content of the solder must not be less than 95 per cent. If platinum is soldered to a carat gold then the appropriate carat-gold solder must be used.

Fluxes

The need to ensure good wettability and bonding by the use of a suitable flux has already been stated. Sodium tetraborate (borax), which is commonly used as a flux for melting silver and the carat golds, itself melts at 760°C which is too high for many of the solders we have considered. Proprietary fluxes containing additives such as potassium fluoride which lower the melting range of the flux are available on the market. Examples are:

Easy-flo flux	Active range 550– 800°C
Tenacity Flux 4A	Active range 600– 850°C
Tenacity Flux 5A	Active range 600–1000°C

Soldering practice

Jewellers and silversmiths traditionally solder using a torch flame. It is necessary to have a good knowledge of the chemistry of the flame in order to achieve a satisfactory joint. The blue cone at the flame centre gives reducing conditions which are necessary to prevent oxide forming on the joint surfaces. The outer zone gives oxidizing conditions and joints will tend to be porous and brittle if this part of the flame is used. Maximum temperatures in the flame are found at the tip of the blue cone.

More recently, a degree of automation has been introduced in the mass-production side of the industry with the use of banks of automated flame burners or electrically-heated conveyor-belt furnaces with protective atmospheres.

The solder may be applied to the joint manually by touching it with a strip or wire of the solder or by placing solder shim stock in the joint prior to heating. Recently, solder pastes which contain the solder in powder form together with a suitable flux and binder have been introduced, particularly for the mass-production of soldered joints, the paste being automatically dispensed from syringes.

With chain making where every link must be soldered after the chain-making process, two techniques are in common use. In the first, the chain is dipped into powdered solder containing a binder and flux such that a sufficient amount remains in the joint gap. The chain is then passed through a flame or furnace to melt the solder. In the other technique wire with a solder core is used in chain making. Subsequent heating causes the solder core to melt and flow across the gap in the links. The solder core is introduced initially down the centre of a hollowed-out cast ingot from which the wire is to be made.

In some products it is necessary to prevent unwanted flow of the solder. This is particularly true in soldering mesh such as watch straps. Traditionally, a paste of jeweller's rouge in water or methylated spirit is applied either side of the joint as a 'stop-off' but a colloidal suspension of graphite in alcohol or a silica sol known as 'Syton 2X' have been found to be superior in preventing oxidation in gold products.

After soldering, flux residues and stop-offs must be removed either by immersion in a suitable pickle or by ultrasonic cleaning.

Welding

In welding, as distinct from soldering, there is fusion of the joint surfaces. Many welding processes incorporate the use of filler metal in the joint gap which melts together with the joint surfaces when heat is applied. Alternatively, the joint may be welded without the use of a filler metal but by applying pressure across the joint as the heat is applied, causing fusion and bonding of the surfaces. Two processes of the latter type are used in the jewellery industry.

Tack welding

This is a form of resistance spot welding where the joint assembly is pressed together lightly between copper electrodes such that the article is not damaged. A pulse of heavy current is passed across the electrodes through the joint, causing fusion at the junction. The process is recommended where it is necessary to hold a complex assembly of pieces in position during soldering.

Electric resistance butt welding

The process is essentially similar to that of tack welding. Low-voltage current is applied to a pair of clamps each holding a piece to be welded with the joint area between them. The two faces are pressed together and they are fused by the passage of the current. Nine-carat gold ring shanks have been welded by this method, although golds containing zinc are not satisfactory for this purpose.

Adhesive joining

Adhesives have been employed for some time in the fashion jewellery industry for attachment of findings and decorative features and for holding stones. In the production of silverware, adhesives are used for fitting non-metallic parts such as wooden and ceramic handles. However, it is only recently that interest has been shown in this method of joining for precious-metal jewellery.

The advantages are that joining is done at room or relatively low temperatures and therefore it is possible to retain the strength introduced by cold-working. Pickling, polishing and texturing procedures are simplified. The disadvantages are that the joints are weaker than hard-soldered joints, curing times may be long and it is difficult to cover glue lines by subsequent electroplating as they are non-conducting. Particular attention must be paid to joint design in order to obtain adequate tensile and shear strength. Certain epoxy-type and cyanoacrylate adhesives are found to be suitable for jewellery applications.

12
Assaying and hallmarking

The legal standards for gold, silver and platinum alloys for jewellery and decorative precious metalware have been given in the relevant chapters. It is now time to look at the methods of assaying and hallmarking that are used to assess these standards.

It is compulsory in the UK to submit articles intended for sale as precious metalware and jewellery to an assay office for analysis and hallmarking; this is not the case for many other countries. This applies to transactions by jewellers, antique dealers, gift shops, auctioneers, pawnbrokers, etc., subject to certain exemptions. It does not apply to single private transactions.

Assaying and marking has been controlled by Act of Parliament since the Middle Ages under the direct supervision of responsible authorities such as a guild of goldsmiths. In London, the responsible authority has been the Worshipful Company of Goldsmiths at the Goldsmiths' Hall. Because the articles were marked at the Hall to show that they were of the legal standard, the word 'hallmark' came into the English language to signify a guarantee of quality.

Interference with the process of hallmarking has always been considered to be a felony and as recently as 1815 was punishable by death. Even today the penalties are very severe. For summary conviction there is a fine not exceeding £1000. For a conviction on indictment the penalty is an unlimited fine or a term of imprisonment not exceeding two years or both, except in the case of forgery or counterfeiting in which case the terms of imprisonment is one not exceeding ten years.

Hallmarking Act 1973

The most recent Act was passed in 1973 with amendments in 1975 and 1976. It is pertinent to give a brief description of the important features of the Act.

1. It is an offence for any person in the course of trade or business to apply to an unhallmarked article a description indicating it is wholly or partly made of gold, silver or platinum, or to supply or offer to supply an unhallmarked article to which such a description has been applied.

2.   The above provisions do not apply to articles described as 'gold-plated', 'rolled gold', 'silver-plated' or 'platinum-plated' provided that the description is true and that if the description is in writing the lettering of 'plated' or 'rolled' is at least as large as any other lettering in the description.

3.   Imported articles are required to be hallmarked, and this is the responsibility of the importer. Marks struck by an authorized assay office in or outside the UK in accordance with the International Convention ratified in 1976 by the EFTA countries are recognized, with the exceptions of 800 and 830 silver which are outside the British legal standards, and it is not necessary to submit these to a UK assay office.

4.   Certain articles are exempt from hallmarking. These include items which are too small and which would be damaged by application of the hallmarking punch, e.g. gold articles weighing less than 1 g, silver less than 5 g and platinum less than 0.5 g, thread and fine chain. However, such items must still be made to a legal standard. A complete list of exemptions is given in the Act and any doubtful cases can be referred to the British Hallmarking Council, PO Box Number 47, Birmingham, B3 2RP. This body was established in 1974 and is charged with the duty of ensuring that adequate facilities for assaying and hallmarking are available in the UK and of ensuring enforcement of the law.

5.   A 'sponsor's mark' (formerly known as the maker's mark) must be stamped on an article. Each assay office keeps a separate register of sponsor's marks registered at that office.

6.   Articles fabricated from separate pieces by the use of solder will not be hallmarked if the use of solder is excessive or if the solder does not comply with the required fineness:

Gold articles: the solder must be of a fineness not less than the standard of fineness of the article, except
   (a) for 22 ct articles, the fineness of solder must not be less than 750 (18 ct).
   (b) for 18 ct filigree work and watch cases, the fineness of solder must not be less than 740, and
   (c) for 18 ct white gold, the fineness of solder must not be less than 500;
Silver articles: the fineness of solder must not be less than 650;
Platinum articles: the solder must be gold, silver, platinum or palladium or a combination of two or more of these metals and a combined fineness of not less than 950.

7.   There are also regulations covering the manufacture of articles made of two or more precious metals, articles made by combining precious metal with other materials such as base metal, ceramics, plastics, etc., and alterations to hallmarked articles.

Further information can be obtained from the British Hallmarking Council or by reference to the Act itself.

Assaying

There are four assay Offices incorporated by Royal Charter or statute and which are independent of any trade organization. These are in London,

Birmingham, Sheffield and Edinburgh. There were other assay offices in former times, notably at Chester, Exeter and Glasgow.

Considering the wide variety of work produced every year ranging from chain, rings, charms, watch cases, brooches, to larger items such as tea and coffee pots, candlesticks, tableware, etc., it is a mammoth task to complete the assaying and marking in a reasonably short time. For instance, the London Assay Office receives about 10 million articles every year and employs a staff of 180 to deal with the work.

Articles must not be incomplete but are submitted at a suitable stage of manufacture so that scraper marks and any distortion caused by marking can be removed during subsequent finishing. In the case of articles made by lost wax casting, sprues should be left on the article so that scrapings may be taken from them without damage to the article. Advice is readily given by the Assay Office on the manner in which articles may be submitted.

Before the assay is made, a representative sample is taken as a small weight of scrapings from the surface of the article or as stated from the sprues of castings. If the article is very intricate, a small sample of the wire or sheet used in manufacture may be included to keep the scrapings to a minimum. If the article is fabricated from a number of pieces, each individual piece is sampled.

The method of assay differs for gold, silver and platinum and so we will consider each in turn.

Gold

Although there are a number of techniques for analysing for gold content, one of the oldest is the fire assay which is extremely accurate and is the method that has been used for the past six centuries to the present. The Romans knew that lead could be used to separate gold and silver from base metal but it was the discovery of nitric acid in the 12th century AD that allowed the fire assay to be used for gold articles.

The scrapings are weighed to an accuracy of one part in 10 000 and wrapped together with a piece of silver in lead foil. The lead foil and its contents are placed in a shallow crater on the top of a porous brick known as a cupel. During cupellation, i.e. a furnace treatment at 1100°C, the lead melts, oxidizes and dissolves out the base-metal content of the scrapings. The oxide melt is absorbed into the porous brick cupel, leaving behind a bead of gold and silver. After removal from the cupellation furnace, the bead is flattened and rolled round into a shape called a cornet. The reason for adding the extra piece of silver to the scrapings is to ensure that the gold can be separated from any silver originally present in the scrapings, a process known as 'parting'. The cornets are placed in a platinum-lined tray and treated with boiling nitric acid which dissolves out the silver leaving only the gold that was present in the original sample. It is necessary to ensure a silver:gold ratio of 2:1 for 'parting' to be successful, hence the need for additional silver. The residual gold is weighed and by comparison with the original weight of scrapings the fineness of gold in the article can be calculated to an accuracy of one part in ten thousand (0.01 per cent).

Silver

For a sterling-silver assay it is assumed until proved otherwise that the sample contains 92.5 per cent silver. A known standard weight of scrapings is placed in a bottle, dissolved in nitric acid and then a measured amount of sodium chloride solution (NaCl) of standard composition is added to precipitate out 92.3 per cent silver as silver chloride, i.e. just enough NaCl solution is added to leave 0.2 per cent Ag in solution. The precipitate is allowed to settle and then a small additional amount of NaCl solution is added to the clear liquid above the precipitate. Provided further precipitation of silver chloride is observed, i.e. the remaining 0.2 per cent Ag now precipitated out, it can be assumed that the original scrapings contained not less than 92.5 per cent Ag. If no further precipitation occurs then the article fails the assay. A similar procedure is adopted for silver of Britannia standard (958.4 fineness).

Platinum

The method used for platinum is atomic absorption spectrophotometry which is a modern technique. Scrapings of about 10 mg are weighed to a high accuracy and dissolved in hot aqua regia for a period of eight to 10 hours. This is made up to a fixed volume in a copper salt solution and placed in the spectrophotometer. The amount of light radiation from a flame absorbed by the platinum in solution is measured and compared with that of a control sample containing exactly 95 per cent platinum. Provided an equivalent amount of radiation is absorbed, the article passes the assay.

Touchstone tests

In addition to the detailed analytical techniques described above, sampling may be done by a touchstone test to indicate the quality of the metal. A touchstone is a smooth hard black stone. The surface to be tested is rubbed on the stone to produce a streak about 10 mm long and 3 mm wide. Acid solutions of different strengths and compositions are applied as a droplet to the streak, and the observed changes such as a colour change or disappearance of the streak indicate the quality of the metal. By selective use of the correct reagents, distinction can be made between the coloured and palladium- and nickel-white golds at different caratages, silver, platinum, palladium and base metal. To assist in identification it is advisable to compare the unknown sample with the rubbing from a 'touch needle' of known appropriate composition. Such tests are of value to jewellers, antique dealers and pawnbrokers in cases where it is not convenient to submit the article to a full assay.

Hallmarking

Articles which have successfully passed the assay are then marked unless they are exempt. The history of hallmarking is fascinating but it is impossible to delve deeply into this in a book of this nature, and collectors and those with a general interest are referred to the bibliography.

Figure 12.1. Example of a hallmark. It shows the sponsor's mark, followed by the English mark for sterling silver, the London Assay Office mark and the date letter for 1976 (reproduced by the permission of the Joint Committee of the Assay Offices of Great Britain. Copyright reserved)

A complete hallmark shows the sponsor's mark followed by the assay mark for the precious-metal alloy concerned, the Assay Office mark and the date letter for the year in which the assay was made (Figure 12.1).

The oldest mark is the leopard's head which was also originally known as the 'King's Mark' and it was introduced in the year 1300 to signify that the metal came up to the required standard. In later times it became the mark of the London Assay Office. Between the years 1478 and 1821 the leopard's head was surmounted by a crown.

Standard Mark

British Articles

Prior to 1975	Standard	From 1975
	22 carat gold Marked in England Marked in Scotland	
	18 carat gold Marked in England Marked in Scotland	
	14 carat gold	
	9 carat gold	
	Sterling silver Marked in England Marked in Scotland	
	Britannia silver	
—	Platinum	

Imported Articles

Prior to 1975		From 1975
	22 carat gold	
	18 carat gold	
	14 carat gold	
	9 carat gold	
	Sterling silver	
	Britannia silver	
—	Platinum	

Assay Office Mark

British Articles

Prior to 1975		Assay Office	From 1975
gold & Sterling silver	Britannia silver	London	gold, silver & platinum
gold	silver	Birmingham	gold & platinum silver
gold	silver	Sheffield	gold, silver & platinum
gold & silver		Edinburgh	gold & silver

Notes – (i) Some variations in the surrounding shields are found before 1975. (ii) All Assay Offices mark Britannia silver, but only London (prior to 1975) had a special Assay Office mark for this standard.

Imported Articles

Prior to 1975		Assay Office	From 1975	
gold	silver		gold & silver	platinum
		London	unchanged	
		Birmingham	unchanged	
		Sheffield	unchanged	
		Edinburgh	unchanged	—

Figure 12.2. Standard and Assay Office marks for British-made and imported articles (reproduced by the permission of the Joint Committee of the Assay Offices of Great Britain. Copyright reserved)

The second mark to be introduced in 1363 was the maker's mark, now the sponsor's mark, which was originally a personal sign or symbol but after 1720 the situation was clarified by the use of the initial letters of the forename and surname.

In 1544, the lion passant was used in place of the leopard's head to indicate the legal standard for both gold and silver ware, this situation remaining until 1798 when a separate mark for 18-carat gold appeared. Its use for 22 ct gold disappeared in 1844 and it is now solely used for silver of sterling quality. Figure 12.2 shows the British marks for both British-made and imported articles.

A letter of the alphabet indicates the year in which the assay was done. At the beginning of each cycle of letters, the script is changed and the outline of the surrounding shield may also be changed. The present cycle started in 1975 to bring all four assay offices into line.

Between 1784 and 1890, a fifth mark, the sovereign's head, was incorporated to show that duty had been paid on the article bearing it as during that period a tax was levied on gold and silver ware. The tax was abolished in 1890. Nowadays, the sovereign's head is used to commemorate important events such as a coronation or silver jubilee (1935, 1953, and 1977).

Fakes and forgeries

In spite of the heavy penalties mentioned earlier, it is not unknown for collectors to come across fakes or forgeries. These may include letting in a piece containing an old hallmark into a modern reproduction; using a hallmarked item such as a piece of cutlery as a pattern for a casting mould to produce identical castings in worthless metal; simply forging old hallmarks onto modern pieces. However, by the use of sophisticated analytical techniques which detect the minute levels of impurities, the assay offices can ascertain within a fairly close period of time when an article was made because improvements in refining gold and silver over the years have meant a reduction in impurity levels.

13
Coatings and surface decoration

Since the earliest times, possession of gold and silver has usually been the privilege of the rich and powerful. Nevertheless, people have always been attracted by the aesthetic beauty of precious metals, particularly gold. Consequently, it is not surprising to find that techniques have been developed by many civilizations to obtain thin coatings of precious metal on a base metal substrate, or on a precious metal substrate containing little or no gold, as a means of achieving an attractive finish on an object at a much lower cost. One such method which has been commercially used since 1840 is electroplating. This will be discussed fully in the next chapter but here we shall explore other types of coatings.

Gilding processes

Gold leaf

The application of thin gold leaf to decorate the surface of any material is one of the oldest crafts, having been in use in many parts of the world from about 4000 BC to the present time. Examples include such diverse objects as death masks, asiatic pagodas and temples, statues, regal crowns, holy figures, altars, mosaics, ceramics, gold edging and lettering on books and shoes, picture frames and so on.

The manufacture of gold leaf is essentially a hand craft. Gold slabs are rolled to about 0.03 mm thickness and cut into small squares known as 'quarters'. These are then interleaved with parchment sheets in a stack which is hammered so that the quarters spread out to the edges of the parchment. The process of quartering and interleaving is repeated to give another stack which is hammered again. Finally, a third stack is prepared for hammering by interleaving with gold beaters skin made from ox intestine. After the final hammering process the thickness of the gold leaf is about 0.0001 mm. It is so thin that it will transmit light with a green colour. It is sold in packets in which each piece of gold leaf is separated by tissue paper.

To achieve this degree of thinness the fineness of gold should be between 920 and 990/1000. Leaf with lower gold contents can be made but the malleability is not so good and it is not so thin.

The application of gold leaf to a surface is a skilled craft. Normally firm pressing and burnishing is sufficient to give an adherent coating but cements such as a solution of albumen (egg white) can be used if required.

Silver leaf can be made by a similar process. However, palladium leaf is often used nowadays instead of silver as it does not tarnish.

Fire gilding

Fire gilding was certainly practised by the ancient Egyptian and Roman civilizations and its use with modifications continued as the main method of gilding up to the advent of electroplating. The cleaned and degreased substrate is first given a dip in a 'quicking' solution of mercuric nitrate and potassium cyanide to prepare the surface giving it a whitish film. A pasty amalgam made from six parts of mercury to one part of gold is smeared over the 'quicked' surface and brushed evenly. Slow heating causes the mercury to evaporate off leaving a bright gold finish. This is a hazardous operation unless adequate safety precautions are taken because of the poisonous nature of mercury vapour. It is no wonder that it has been replaced by gold electroplating.

Water gilding

This is another old process in which the base-metal substrate is dipped in a solution containing a gold salt. Gold is deposited onto the substrate as a form of electroless plate. Such coatings are extremely thin and soon wear off with rubbing when the gilded articles are used.

Depletion gilding

The process was widely practised by the Egyptians and Romans, as described by Pliny and others around the 1st century AD, and also by the pre-Columbian cultures of Central and South America. In principle, alloys containing relatively low amounts of gold are treated in such a manner as to remove the base-metal and silver content from the surface layers, leaving them considerably enriched in gold. The brilliance of the coating may be improved by burnishing. The process is similar to that of 'parting' in which silver is removed from gold by treatment with nitric acid. However, nitric acid was not used before the 12th century and these early civilizations employed other substances such as plant juices containing oxalic acid, urine, common salt (NaCl) and alum. The alloy to be gilded was usually placed in an earthenware crucible and surrounded by a powdered mixture of salts and brick dust. The crucible and contents were heated and the molten salt mixture reacted with the surface of the alloy to form chlorides of silver, copper and other impurity metals. These chlorides were absorbed by the brick dust. After cooling and washing the enriched surface was then burnished to consolidate it.

The pre-Columbians used two types of alloy for depletion gilding. One type is known as 'tumbaga', a reddish-bronze-coloured copper alloy with different gold contents and probably a small amount of silver. The other type were greenish-white ternary silver-gold-copper alloys with a high proportion of silver. These are similar in composition to the 'electrum' of the Mediterranean World.

Sheffield Plate

In 1743 a Sheffield cutler, Thomas Bolsover, discovered a method of fusing sheets of sterling silver to the surfaces of a copper ingot. The copper ingot surfaces were planed and filed. Sheets of sterling silver, similarly prepared, were placed on one or on either side of the ingot, borax was applied to the edges and the whole assembly bound tightly together with wire. This was placed in a coke furnace until diffusion lowered the melting point at the interfaces between the silver and the copper and the silver began to 'weep'. After removal from the furnace, cooling and pickling, the coated ingot was rolled to give copper sheet clad with a thin layer of sterling silver for subsequent manufacture into flatware and holloware.

This was very popular with the middle classes because it enabled them to buy a susbtitute form of silverware at prices they could afford. Today, Sheffield Plate has considerable antique value. A system of marks, similar to the maker's marks in hallmarking, was used which is of great assistance to collectors. Sheffield Plate can often be recognized by the underlying copper showing through on high spots due to the silver having been rubbed away by zealous housemaids. Caution must be exercised when collecting, as a similar effect can be found with electroplated silver on copper. One should look carefully at soldered joints, because hard silver solder or soft solder would have been used, and at edges where a length of silver wire was soldered on to conceal the exposed copper.

The advent of electroplating spelt the death knell of the Sheffield Plate industry in the latter half of the 19th century.

Rolled gold

Not long after the invention of Sheffield Plate the first patent (1785) for rolled gold was taken out. The process is very similar, namely, carat gold sheets are sweated or soldered onto clean surfaces of a base-metal ingot and the composite rolled or drawn down to produce sheet, tube or wire. Great care is taken in processing to maintain the integrity of the gold coating. The coating sheet may be either single surface or double surface. A wide variety of finishes, caratages and coating thicknesses are produced. Typically, coating thicknesses range from $\frac{1}{3}$ to $\frac{1}{300}$ expressed as a ratio of coating thickness to total thickness. The choice of thickness, colour and caratage determines the quality and price range of the finished article and its wear and tarnish resistance. The base metal can be gilding metal (e.g. 90 per cent copper—10 per cent zinc), bronze or nickel silver. Rolled-gold coatings on sterling silver are also produced, in which case the articles can be given a silver hallmark.

Certain precautions have to be taken in making articles from semi-fabricated rolled-gold products. Excessive amounts of cold work can lead to cracking or rupture of the coating and excessively high annealing temperatures can lead to exaggerated grain growth and the orange-peel effect. Annealing at 600°C is recommended for most grades of rolled gold.

Granulation

Granulation is the technique of surface ornamentation in which granules, made by melting filings or short pieces of fine wire in layers of charcoal, are fused onto the surface of gold or silver. Items have been found dating from 2500 BC but it was the Etruscans who developed the technique to a high degree of perfection. It is generally assumed that fusion was accomplished without soldering but Jochem Wolters has recently shown that there are many examples where metallic solder has been used. It is probable that a number of differing techniques have been employed over the centuries. One ancient technique thought to have been used by the Etruscans depends on a type of non-metallic solder.

The granules are placed in position on the parent metal surface with an organic adhesive, e.g. gum tragacanth, admixed with a copper compound such as malachite which is a basic copper carbonate. Heating first causes decomposition of the carbonate to cupric oxide at about 100°C and then carbonization of the adhesive at about 600°C. The carbon reduces the copper oxide to metallic copper which then diffuses into the granule and the parent surface lowering the melting point at the interface. Fusion is therefore achieved by alloying.

Many present-day jewellers are using granulation as a form of surface decoration (Figure 13.1).

Niello

This is a form of inlay in silver which goes back to Roman times and is still practised in the Middle and Far East. Chased or engraved patterns on silver are filled with a powdered mixture of silver, copper, lead and sulphur which is then melted and later polished to give a permanent grey-black inlay.

'Oxidized' silver

This is a surface finish, ranging from red to bluish-black in colour, occasionally given to sterling silver objects. The name is a misnomer since the colour is due to the formation of silver and copper sulphides and it is not an oxide film. The coating is obtained by exposing the silver surface to a sulphur compound such as ammonium sulphide or by heating with sulphur in a closed container. The colour of the finish is mainly dependent on the coating thickness.

Enamelling

One very important means of applying surface decoration to precious metalware and jewellery is that of enamelling. There are a number of different enamelling techniques, e.g. cloisonné, plique à jour, basse-taille and champlevé (Figure 13.2).

In essence, the enamel which is normally supplied as lumps of coloured glass of special compositions, is prepared in powder form by grinding the

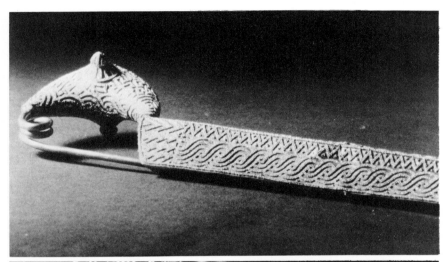

Figure 13.1. An example of granulation (courtesy Pärn Taimsalu, International Gold Corporation)

constituents in a pestle and mortar and then made into a slurry with a suitable liquid so that it can be easily transferred to the surface of the object being enamelled. On heating in a furnace, the liquid evaporates and the powder enamel melts and flows over the area to be covered. Enamels may be considered as relatively low melting-range glasses. Their colour, melting range and transparency or opacity are controlled by the constituents used. Enamelling is a highly skilled craft.

Figure 13.2. Types of enamelling. (a) Champlevé: cells cut into metal and filled with opaque enamel. (b) Basse taille: translucent enamel overlaying chased or engraved surface.
(c) Cloisonné: opaque enamel between narrow strips of metal (cloisons) soldered onto a groundplate. (d) Plique-à-jour: translucent enamel filled in spaces between metal cloisons. The effect is that of a miniature stained-glass window

Attention must also be given to the quality of the metal to be enamelled, including the use of any solder, if enamelling is to be successful. Special grades of carat golds and solders are recommended. For example, zinc contents in carat golds must be kept to low levels.

Japanese decorative metalware

A brief mention must be made of the Japanese craft of 'mokumé'. Here, thin plates of various metals – copper, gold, silver and alloys of these metals – are stacked, heated and sweated together to produce a diffusion bond between each layer. The resulting laminate consists of alternate layers of differing-coloured metals. It is cut or drilled in such a manner so that on subsequent rolling or hammering or twisting, contoured patterns revealing the different-coloured layers are displayed on the surface rather like wood grain. Part of the artistic skill comes in deciding where and how deep the incisions should be made and how the subsequent deformation is done.

Apart from pure metals, two types of alloys were favoured:

Shakudo – this is copper containing 0.5 to 4 per cent gold with traces of silver,

Shibu-ichi – a range of copper-silver alloys (49 to 86 per cent copper–13 to 51 per cent silver with traces of gold) which have a fine silver-grey colour particularly at the higher silver contents.

The other important feature of mokumé is that immersion in various solutions produces rich patinas, the colour depending on the composition of the solution. For example, a beautiful purple colour can be obtained on shakudo with a solution of copper sulphate, verdigris and water.

These patinated alloys were not only used for mokumé work but also for ornamental samurai-sword furniture.

14
Electroplating and allied processes

In the previous chapter, various methods of producing decorative precious metal finishes on base-metal substrates were discussed. Today, these have largely been superseded by electroplating, with the notable exceptions of rolled gold and gold-leaf gilding of non-metallic materials and very large objects. Electroplating is the application of electrodeposition to produce a thin adherent solid metal coating on a substrate.

Although there is evidence to suggest that some form of wet-cell battery existed in Persia over 2000 years ago and may have been used for some form of gold electroplating, it was the invention of the battery by Count Alessandro Volta (Volta's pile) at the end of the 18th century which led directly to the process of electroplating. Brugnatelli, an Italian Professor of Chemistry, gave an account of silver plating in 1800 and of gold plating in 1805 in which he used Volta's pile. However, the commercial development of electroplating was started by the work of the Elkington Brothers in Birmingham and John Wright, a surgeon, who were granted a patent in 1840 for the electrodeposition of silver.

In the jewellery and allied trades, the use of precious-metal plating is to

1. prevent the corrosion of the underlying metallic substrate,
2. produce tarnish-resistant coatings by the use of gold and platinum-group metal plating,
3. improve wear resistance, and
4. give a highly-reflective decorative finish.

In addition, the process is suited to large production runs of small articles.

A variety of substrates can be used, the most common being steel, zinc die-casting alloys, lead-tin alloys and copper and copper-base alloys such as gilding metal and nickel silver. Much of the costume jewellery worn at the present time is lead-tin alloy which, because of its low melting range, can be conveniently centrifugally-cast in vulcanized rubber moulds and subsequently gold-plated.

The Elkington process initially used copper as the substrate, which is why collectors of Sheffield Plate have to be careful not to be confused by articles produced by electroplating. Within a short time the use of copper was largely replaced by that of nickel silver, hence the term 'electroplated nickel silver' (EPNS) for good-quality silver plate. Nickel silver, also called German Silver as most of it was originally exported from Germany in the latter half of the 19th century, is a misnomer because it contains no silver. Nickel silvers are Cu–Zn–Ni alloys and have characteristics similar to those of the brasses (Cu–Zn alloys). The presence of nickel, together with zinc, has a pronounced whitening effect. Compositions vary widely with nickel from 7 to 35 per cent, zinc from 10 to 35 per cent and copper from 50 to 65 per cent. Typically, an alloy containing about 18 per cent Ni is best suited for flatware and holloware because it is amenable to the same sheet-metal working processes as used for sterling silver, and it is sufficiently white not to look too unsightly if the silver plate is worn away on the high spots.

Non-metallic substrates may be plated provided their surfaces are made electrically conducting. This can be achieved by silver spraying, metal sputtering or the application of graphite or flake metal powder onto the surface prior to plating.

Principles of electroplating

The principles of electroplating are in essence the same as those for electrorefining, which was briefly discussed in Chapter 2, although the process conditions are very different. In electrorefining the metal is deposited in large quantity on cathodes, and these will subsequently be stripped off and melted down. Apart from ensuring purity, the quality of the deposited coating is secondary. In electroplating, strict quality control is of prime importance and the whole process, including pre- and post-treatments, is closely monitored.

The articles to be plated are suspended in a plating bath and connected to the negative terminal of a DC supply, i.e. the cathode. Anodes are connected to the positive terminal and are also suspended in the bath (Figure 14.1). The DC supply is nowadays obtained via a transformer rectifier from the mains AC supply. An ammeter to measure electrical current, a variable resistor to control the current, and a voltmeter to measure the electrode potential, complete the electrical circuitry (Figure 14.2).

The plating bath is a vat containing an electrolytically-conducting plating solution known as the electrolyte. Modern plating solutions may be very complex with additives for various aspects of plating control, but essentially they contain a salt of the metal to be deposited.

The anodes are of two types. With soluble anodes, which are made of the same metal as that being plated, the metal is transferred via the plating solution and re-deposited on the article being plated at the cathode. Obviously, under these circumstances the anodes eventually have to be replaced.

Insoluble anodes are not dissolved in the electrolyte. They are usually made from stainless steel but graphite, platinum or its cheaper substitute, platinized titanium, may be used as alternatives. Here, the metal being deposited is gradually depleted from the plating solution and its composition has to be maintained by topping up the solution periodically with fresh metallic salt.

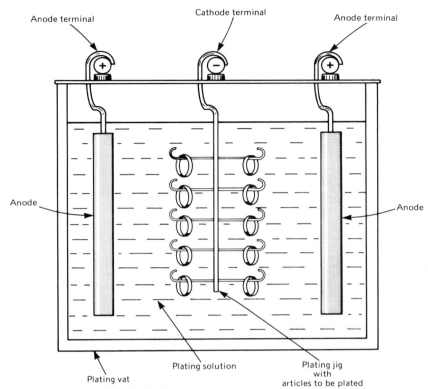

Figure 14.1. A typical plating bath

The electrochemistry of plating is far too complicated to deal with here, and only a brief simplified version of the mechanism is possible. In solution the metallic salt ionizes, i.e. it splits into a part containing the metal, which may be a positively-charged cation or a complex ion, and a part which is negatively charged and known as the anion. In the presence of an electric field, the ions containing the metal migrate to the cathode where they are reduced to give solid metal which is deposited as a layer on the cathode surface. The anions migrate to the positively-charged anode. If a soluble anode is used, metal from the anode reacts with the anion to form the salt and the process of ionization and migration continues.

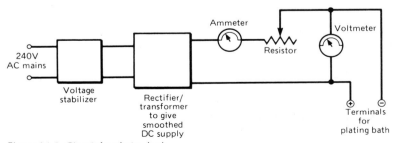

Figure 14.2. Circuit for plating bath

The electrochemistry of the process in part determines the quality of the coating. For example, silver nitrate $AgNO_3$ ionizes and the Ag^+ cation will migrate to the cathode but a very spongy deposit is obtained which is easily rubbed off. A fine-grained smooth adherent deposit is required and this dictates the composition of the electrolyte. In the case of both gold and silver plating this necessitates the use of cyanide salts (AuCN and AgCN).

Other additions

Commercial plating solutions contain a number of other salts and compounds in addition to those of the metal to be plated. These are added to promote certain characteristics to the coating and they include grain refiners, brighteners and levellers to improve the adherence, smoothness and reflectivity of the coating; stress reducers which (as the name implies) lower the level of internal stress in the coating; agents to improve 'throwing power' to ensure a more uniform thickness of deposit on a complex shaped article; anti-pitting agents; additions to control conductivity and the pH of the solution. The pH of an aqueous solution can be regarded as a measure of its degree of acidity or alkalinity. The value of pH for a neutral solution such as pure water is 7. Values lower than 7 indicate increasing acidity, and greater than 7 increasing alkalinity.

There are many proprietary plating solutions on the market, the compositions of which are a trade secret. For anyone intending to use the process it is advisable to purchase these in preference to trying to make up a solution.

Operating conditions

The operating voltage depends on the composition of the electrolyte but is typically in the range 0.5–7 volts.

The current determines the rate at which the metal is deposited. This can be calculated. For example, a current of one ampere deposits 4.02 g of silver, 2.45 g of gold and 1.82 g of platinum per hour. Faster rates of deposition can be obtained with higher currents but this may be at the expense of coating quality and the optimum level has to be found. Obviously, the cathode surface area being plated is important since this determines the thickness of coating for a given current. For this reason, it is more sensible to express the required current for good-quality plating in terms of the 'current density' in amps per square metre of cathode surface.

The temperature of the plating bath is another important variable which needs to be controlled, in some cases up to 80°C. Modern installations include immersion heaters for this purpose.

The items to be plated should be immersed in the plating solution, either by using soft copper wire or by placing them on special plating jigs to avoid damaging the coating. The contact point needs to be changed occasionally, perhaps by periodic shaking, otherwise this would be a weak part of the coating.

For a large production run of very small items it is not convenient to wire up or use a plating jig. In these circumstances, they are tumbled in a plating barrel which contains the electrolyte. The wall of the barrel is the cathode and electrical contact with the articles is maintained during the tumbling action. The axle on which the barrel turns is the anode which is electrically insulated from the rest of the barrel.

Pre- and post-plating treatments

Good surface preparation is vital to the success of the electroplating process. Dust particles and scratches will show up unless the plate is very thick. Therefore, a dust-free environment well away from buffing and polishing equipment is advised and a reasonable degree of polishing to remove scratches is required. Grease interferes with the adherence of the plate as will oxide films. Articles may be treated initially in a vapour degreasing tank containing a hot solvent such as trichloroethylene. They are then usually electrolytically degreased in a trisodium phosphate solution. Hydrogen evolution on the surface of the cathodic articles assists in the removal of grease films. Electrolytic acid pickling may be used to remove base-metal-oxide films. Ultrasonic cleaning baths, in which the ultrasonic waves break up and remove surface films, have found favour recently.

After each stage of the pre-treatment, thorough rinsing is essential. The presence of dewetting on removal from the rinse bath reveals signs of inadequate degreasing.

On completion of the plating process, the parts are again thoroughly rinsed and then dried in sawdust or a centrifugal drier.

Precious-metal plating practice

Silver plating

Cyanide salt baths are used almost exclusively with fine silver soluble anodes in sheet form which themselves should have a similar surface area to that of the work to be plated. The bath is normally operated at room temperature. A typical operating voltage is 0.5–1.5 volts and the current density is of the order of $60 \, A \, m^{-2}$.

To improve the adherence of the deposit, it is necessary to apply a preliminary coating. This used to be done by dipping in a solution containing mercury salts to produce a thin mercury film which amalgamates with both the base metal substrate and the silver plate. This was known as 'quicking'. Nowadays the 'silver strike' is applied electrolytically in a potassium silver-cyanide solution operative at 2–3 volts and a high-current density of at least $100 \, A \, m^{-2}$ for about 30 seconds prior to actual plating.

Plating thickness depends on operating conditions and plating time. A thickness of 0.015 mm (15 microns) may be regarded as suitable for general work, while as much as 0.09–0.1 mm may be applied for the best-quality plate.

Particular mention should be made of the practice of 'bright silver plating'. A number of years ago it was discovered that a small addition of a stock solution of carbon disulphide to the plating bath gave a deposit which was much brighter in appearance compared to that normally obtained and which was harder, giving improved wear-resistance. The use of carbon disulphide has been superseded by brighteners based on complex organic sulphur compounds.

The finish is achieved by the restriction of crystal growth during formation of the deposit, giving a compact smooth bright plate. In addition, the presence of the 'locked-in' sulphur compounds blocks the active silver centres from attack by hydrogen sulphide imparting a degree of tarnish-resistance. Even during wear caused by rubbing, more inhibitor is exposed and the tarnish-delaying properties are retained.

Plated articles are subjected to final finishing treatments to achieve different lustres, e.g. polished, frosted, satin and butler finish.

Gold plating

During the last 30 years a revolution has taken place in the gold electroplating industry. In former times, the electrolytes were based on hot alkaline cyanide solutions with bath temperatures in the range 50–75°C and current densities of 20–60 $A\,m^{-2}$. The problem with these solutions was that the gold coatings were thick, coarse-grained and uneven and required scratch brushing, burnishing and polishing to obtain a bright finish. During the 1950s a need arose in the electronics industry for hard thin gold deposits on components for improved wear- and corrosion-resistance. This led to the development of bright gold plating from acid cyanide solutions. The process may be operated either at room temperature or using hot solutions with current densities in the range 10–100 $A\,m^{-2}$. The coatings are fine-grained and smooth, harder, more wear-resistant and less porous, thus obviating the need for costly burnishing and polishing treatments.

Generally plate thickness is in the range 0.5 to 2.5 microns, but for some applications, notably electronic components, 'heavy gold plating' with thicknesses in the range 2.5–25 microns are common. The term 'electrogilding' usually refers to flash deposits having thicknesses 0.25–0.5 microns.

The caratage of the gold plate can be altered by adjustment of the electrolyte composition and operating conditions. Additions of silver and copper cyanides and salts of other metals such as nickel, cobalt and cadmium will give a wide range of caratage (9 to 24 ct), colour and finish.

Soluble gold anodes may be used but are too easily stolen and so it is general practice to use insoluble anodes and maintain bath composition by periodically adding more plating salts.

Platinum-group metal plating

The three members of this group which are plated commercially are platinum, palladium and rhodium. Reference has already been made to the use of rhodium plating on silverware as a means of providing tarnish resistance. The cyanide salts of the platinum-group metals are too stable for plating but suitable plating solutions are available on the market.

Electroforming

Electroforming is the production of articles by electrodeposition where the completed article is removed from the substrate on which it has been deposited.

In principle the process is the same as that of electroplating, the essential difference being that the current densities are very high, e.g. $50-350\,A\,m^{-2}$ to give high rates of deposition and to enable very thick layers to be produced in just a few hours. Thicknesses of $0.1-0.35\,mm$ ($100-350$ microns) have been reported for items such as earrings, pendants and bracelets in 18 ct gold, although much greater thicknesses can be produced

Figure 14.3. Coronet made by Louis Osman for the investiture of the Prince of Wales in 1969 (reproduced by gracious permission of Her Majesty the Queen)

provided the deposition time is extended. Special plating solutions are recommended where electroforming is to be used. Another feature of electroformed articles are that they are very fine-grained and much harder than can be obtained by working or casting metal of the same purity. This is obviously an advantage for items which are relatively thin but have to be rigid and self-supporting. The process is also suitable for silver articles.

One notable piece of work which was made by this process is the coronet designed by Louis Osman for the investiture of the Prince of Wales at Caernarvon Castle in 1969 (Figure 14.3).

Provided the surface of the substrate can be made electrically conducting, an electroform can be built up on virtually any object. Replicas of coins, medals and flat objects are formed on silicone rubber or PVC negative moulds which give excellent reproduction of detail from the original master pattern. Epoxy resins can be used where more rigid moulds are required. Hollow three-dimensional forms are produced on wax models, similar to those used for lost-wax casting, or with low-melting 'Cerro' alloys (tin-bismuth m.p. 120°C) which can be melted out after electrodeposition. For non-metallic moulds and mandrels, the surfaces are sprayed with ammoniacal silver nitrate solution with a reducing agent, e.g. formaldehyde, through a twin-jet gun. A thin bright coherent conductive coating of silver is produced to which electrical contact can be made.

Electropolishing

Electropolishing is the reverse of electroplating in that the articles to be polished are made the anodes and metal is electrolytically removed from their surfaces. This is deposited on insoluble cathodes such as stainless steel where it may easily be recovered.

In mechanical polishing, burnishing and fine abrasion removes the high spots (asperities) and fills in any microcavities to give a smooth bright surface. During electropolishing, the asperities are dissolved away preferentially until a smooth finish is obtained. The operating conditions of the bath, i.e. current density, voltage, temperature, electrolyte composition, etc., have to be adjusted to give optimum polishing conditions, otherwise electrolytic etching and electrostripping may occur (Figure 14.4).

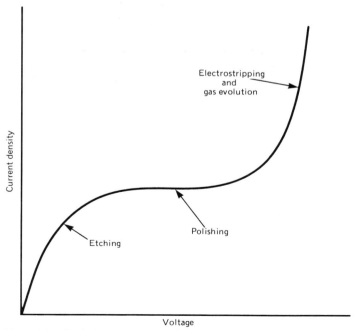

Figure 14.4. Typical current versus voltage curve for determining polishing conditions

The process is widely used as a finishing treatment for gold jewellery, particularly for complex shapes such as investment castings, where mechanical polishing would be difficult. Cyanide solutions are used at 70–80°C at high-current densities of the order of $1000\,A\,m^{-2}$ for a few minutes. Although silver can be electropolished, this is not commonly done as a finishing process in the silverware trade apart from small items of jewellery.

Electrostripping

Electrostripping is an extension of electropolishing in that the operating conditions are such that material is rapidly removed (Figure 14.4). It is employed in the silver-plating trade when an unsatisfactory deposit of silver plate has been obtained and it is necessary to strip this off before a fresh start can be made. It may also be used to strip off unsightly areas of firestain from sterling silverware prior to final polishing.

A warning

The cyanide solutions used for the processes described in this chapter are highly poisonous. Everyone is aware of the fatal results if cyanide is swallowed. What may not be so well known is that if acid comes into contact with cyanide, deadly fumes of hydrocyanic acid (prussic acid) are released. It is vital that thorough rinsing of acid-pickled objects is done to avoid carrying over trapped pockets of residual acid into the plating and polishing baths. Antidotes must always be near at hand when cyanides are used.

It is equally important that cyanides are not washed away down the drains. Expert advice on effluent disposal of cyanide solutions and residues, including rinses, should be sought before any of these processes are used.

15
Other uses of precious metals

In a book of this nature, which is largely intended as an introduction to the precious metals, it is impossible to deal with all the important aspects of these metals and we have concentrated on their use for jewellery and decorative metalware. This chapter is written in an attempt to redress the balance to some extent by noting the role played by the precious metals in other important areas, particularly in today's technological society.

Before dealing with the uses of the individual metals, we shall look at two areas where most of them have made a contribution, namely, coinage and dental alloys.

Coinage

From the earliest days of their availability, a major use of gold and silver has been a monetary one because of their indestructability and the fact that they offered a safe store of value and an easy means of exchange. Originally, the monetary value was solely based on weight, e.g. the Hebrew 'shekel' weighed the equivalent of about ½ oz. The Lydians in the late 7th century BC were the first to introduce coins of electrum exhibiting a regular shape and which were stamped with marks indicating weight and an official authenticating sign vested in a public authority. This region had rich gold deposits and it was from here that the legend of King Midas and his golden touch originated. King Croessus of Lydia, famed for his fabulous wealth, was the first man to introduce a bimetallic currency standard for gold and silver coinage using a fixed gold coin:silver coin ratio. The Persian ruler, Darius, introduced the 'daric' about 450 BC which enjoyed a high reputation because it contained 97 per cent gold.

The Romans adopted gold and silver for their monetary system in 269 BC, the 'solidus' and 'denarius', respectively. They also fixed the gold:silver ratio at 1:10 as a means for restricting the minting of silver coins and restraining inflation. The use of a ratio persisted through to about 1865 when it was 1:15½. Generally, three classes of coins were in use: gold coins for governments and the wealthy, silver coins for merchants and trade and copper alloy coinage for everyday needs.

In more recent times, Henry VII introduced the gold sovereign in 1489 and it weighed 15.6 g, but the first modern gold sovereign, as we know it today, was minted in 1817 in the reign of George III as a result of a new coinage system. Other gold coins, such as the 'spade guinea', were also in circulation at this time. The gold sovereign is 22 ct gold (916 fineness) and weighs 7.98 g.

In 1823, Britain was the first country to adopt the Gold Standard, and this was quickly supported by other major countries. Under this system, paper money could be exchanged for its face value with gold. It was also a means of regulating international trade. The Gold Standard collapsed in 1914 with the advent of the First World War and attempts to revive it proved unsuccessful.

The use of sterling silver in the UK for silver coinage was discontinued in 1920 and replaced with an alloy containing 50 per cent silver. In 1946, a white cupro-nickel alloy was introduced for 'silver coinage' and the use of silver ceased for legal tender. With the demonetarization of silver, its price fell to its lowest level for over 100 years in 1931 (5p an oz) but the increasing use of silver for industrial purposes, as well as inflation, has since forced the price up by a considerable amount. Today, silver is used for proof coinage, commemorative coins, medals and plaques.

A remarkable change has recently taken place in gold-coin production. From 1969 to 1979 the amount of gold used for coinage has increased from 832 000 oz to 9 900 000 oz. This is due to the introduction of the bullion coin which answers the problem posed by a fluctuating gold price and a fixed value. Bullion coin is not related to an official currency. The well-known South African 'Krugerrand' contains one troy oz of pure gold minted as a 22 ct gold-copper alloy. The Canadian 'Maple Leaf' is minted from 1 oz of fine gold and in 1975 Russia introduced the 'Chervonetz' containing $\frac{1}{4}$ oz (7.78 g) of fine gold. In addition, during this period, the number of nations minting 22 ct legal-tender gold coins has risen from five to 80.

So far, no mention has been made of platinum. A very small percentage of platinum is used today for proof coins and for commemorative purposes, but between 1828 and 1846 Russia minted a large number of platinum coins as part of its official currency in order to utilize the output of its recently-discovered mines. In 1983, a 1 oz bullion coin called a platinum noble was introduced by the Isle of Man.

Dental alloys

Most of us are familiar with the silvery-white amalgams used for filling cavities after tooth decay although it is said that its use is in decline because of the advances made in preventative dental care. The dental alloy, having a composition of 67−70 per cent Ag, 25−28 per cent Sn (tin), 0−5 per cent Cu and 1−2 per cent Zn, is prepared by casting ingots and crushing and grinding them to powder. The composition is such that a brittle intermetallic compound is produced. Shrinkage after subsequent amalgamation and curing is minimal. The next stage takes place when the patient is in the dentist's chair. The powder is mixed with mercury in a definite ratio of five parts of alloy powder to eight parts of mercury by weight and the mixture is quickly put into the prepared cavity where amalgamation and setting takes place.

There is evidence that gold has been used for dentistry since 2500 BC. The Romans practised gold dentistry, although often this was purely for reasons of vanity.

Fine gold is used in three forms: as gold foil about 0.06 mm thick, as powder, and as 'mat gold'. Mat gold is in the form of thin flake-like crystals produced by electrodeposition. The foil is wrapped into pellets or around powder or mat gold, annealed by gentle heating to about 300°C and then cold-pressure-welded into prepared cavities as gold fillings. A gold–40 per cent platinum foil prepared by a sandwich technique is also used for this purpose. It is stronger but more difficult to manipulate.

Gold is also used as a major constituent in dental alloys together with silver, platinum, palladium and base metals. There are three classes of dental alloys: (a) casting alloys, (b) wrought wire and (c) solder alloys. Their compositions are based on the need to have a high resistance to corrosion and tarnishing in the oral environment coupled with good mechanical properties and resistance to rubbing wear.

The casting alloys are used for inlays, bridges and crowns and can be produced with a range of colour depending on alloy composition. Castings are made by the lost-wax process using the plaster models prepared by the dentist. It was the dental profession that 're-discovered' the lost-wax process early this century. The casting alloys are classified according to their hardness. Table 15.1 gives typical compositions. The stronger alloys are age-hardening, giving the strength required to withstand the high pressures developed during eating.

Table 15.1. Compositions of dental gold casting alloys

Type	Au	Ag	Cu	Pd	Pt	Zn
Yellow golds						
Soft	79–92.5	2–3	2–4.5	0.5	0.5	0.5
Medium	75–78	12–14.5	7–10	1–4	1	0.5
Hard	62–78	8–26	8–11	2–4	3	1
Partial denture	60–71.5	4.5–20	11–16	5	8.5	1–2
White golds						
Hard	65–70	7–12	6–10	10–12	4	1–2
Partial denture	28–30	25–30	20–25	15–20	3--7	0.5–1.5

The wrought alloys used in wire form for clasps and orthodontic appliances contain less than 60 per cent gold, but the palladium and silver contents are higher than for the majority of the cast alloys. The dental solders are very similar to the jewellery solders, based on the Au–Ag–Cu system, with 2–3 per cent Sn and 2–4 per cent Zn added to lower the melting temperature.

Since the mid 1950s, the application of cast dental alloys bonded to dental porcelain has been widely adopted (Figure 15.1). The alloys are based on Au–Pd–Pt with up to 1 per cent iron (Fe), indium (In) and tin (Sn) for age-hardening. The presence of platinum and palladium raises the melting range and imparts high-temperature strength to the casting during fusion to the porcelain. Obviously, compatibility of the alloys with dental porcelain is of prime importance. Factors which have to be taken into consideration include the differences in thermal expansion, the glass transition temperature of the porcelain, discolouration of the porcelain during firing (the presence of copper in the alloy should be avoided), and the introduction of residual stresses.

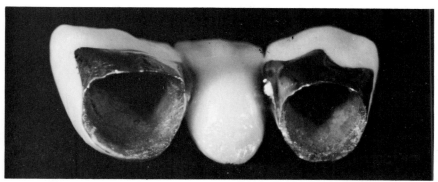

Figure 15.1. Cast dental alloy bonded to porcelain in a three-unit bridge restoration (courtesy A. T. Durham, Engelhard Industries Ltd.)

The escalating price of gold has led to the further development of low gold or palladium-silver or even silver-free palladium alloy compositions. In spite of this, the present annual world consumption of gold for dentistry is 2 600 000 oz.

Other uses of gold

Gold possesses a number of physical, chemical and biochemical properties which are either unique to it or are shared only by the other precious metals. The most important ones are resistance to corrosion and tarnishing, ease of fabrication and high electrical and thermal conductivity.

One major area of use is in the electrical, electronic and telecommunication industries which currently accounts for 70–80 tonnes of gold annually. The emphasis here is on long-term reliability with respect to electrical conductivity, bondability and contact resistance in such applications as computers, calculators, telephone systems, radio, television, and control systems in missiles and spacecraft. Most of these applications involve the use of printed-circuit boards, fine gold wire connectors, relays, gold-based solders and 'gold-doped' silicon semiconductors. Considerable advances have been made in bright gold electroplating and 'thick-film' technology, in which gold coatings are evaporated or sputtered onto a substrate such as a ceramic.

Gold is widely used for evaporated thin films on window glass. These films have the property of reflecting infra-red radiation, i.e. heat rays, while transmitting visible radiation. The advantage is that air-conditioning costs in office buildings are reduced in summer months, and central heating costs are reduced during the winter. The windscreens of locomotives, aircraft and other vehicles may be heated by passing an electric current through a thin gold film on the screen, thereby preventing misting and icing.

In the field of chemical engineering, gold–platinum–rhodium alloy linings are used for vessels where chemical reactions involving highly corrosive materials take place. Gold-alloy spinnerets are employed in synthetic-fibre manufacture. Liquid ammonia tanks have bursting discs which are designed to fail at relatively low pressures. Gold discs are chosen because they have low strength and an indefinitely long life, and they maintain their calibration compared with discs of other materials.

Gold is used in temperature measurement as a thermocouple material. Briefly, a thermocouple consists of two dissimilar metal or alloy wires joined at both ends. If a temperature difference exists between each junction, a current flows around the circuit and an electromotive force can be measured in millivolts. The magnitude of the voltage depends on the temperature difference, and hence the couple can be calibrated for temperature measurement. Normally, a suitable measuring instrument is placed across one junction kept at ambient temperature (20–25°C) and the other junction is placed in the environment at a different temperature, e.g. a furnace or refrigeration plant. Gold-manganese couples are used for the measurement of temperatures at or near absolute zero (−273°C).

Gold-alloy wires are utilized in measuring instruments such as potentiometers and ultra-high pressure gauges. One interesting application is the mercury-detection gauge where the absorption of traces of mercury vapour increases the electrical resistivity of a gold film.

In the medical field, radio-isotope gold-grain implants are used for the treatment of malignant growths. Radioactive gold in colloidal form delineates organs in the body, particularly the lungs and liver, by a technique known as scintography. Gold salts find application in the treatment of rheumatoid arthritis, although their toxic effects mean that their use must be carefully controlled.

It has been estimated that the total industrial use of gold apart from jewellery, coinage and investment bars is 230 tonnes annually.

Other uses of silver

Silver is the best electrical conductor of all metals, and hence it finds wide use for electrical and electronic components, particularly for low- and medium-current switching devices. Compared with copper it has good resistance to oxidation, low contact resistance and high current-carrying capacity. It is estimated that about 10 per cent of silver production is for contact purposes either as fine silver, silver–copper and other alloys or as composites for increased strength. Silver paints are applied as electrically-conductive coatings on ceramics.

Since the 1940s, a range of silver batteries has been developed, although Volta first demonstrated the principle of the battery using silver–zinc alloys as long ago as 1880. Silver–zinc and silver–cadmium secondary batteries deliver a higher energy output per unit weight and space compared with other well-known battery systems, but have a relatively short life and higher manufacturing cost. Consequently, they are mainly used for military and space applications. Secondary batteries are capable of many cycles of discharge, whereas a primary battery can only be used once. Silver chloride—magnesium primary batteries are used as power supplies for torpedoes and sonar-buoys.

In chemical engineering, thin silver bursting discs are employed for low-pressure applications. Silver catalysts are used for oxidation reactions, e.g. the production of formaldehyde and ethylene glycol (anti-freeze). To increase catalytic activity, a large silver contact surface area is desired and this is achieved with wire-mesh screens or silver crystals.

The photographic industry accounts for about 30 per cent of total annual silver production. Emulsions are prepared from a soluble silver salt, e.g. silver nitrate, combined with a soluble alkali halide, such as potassium bromide, to precipitate the silver halide as a fine suspension. A colloidal substance is present to prevent coalescence. A high degree of process control is necessary to produce a range of emulsions for a variety of photographic applications.

In the 14th century, mirrors were first produced by brushing a liquid tin—silver amalgam onto glass and then heating to drive off the mercury, leaving a metallic film. This highly dangerous process was superseded in 1855 when it was discovered that the reduction of silver-nitrate solution by tartaric acid produced metallic silver which could be deposited as a mirror film. Today, this process is highly mechanized. After deposition of the silver, an overlay of copper is electroplated on, and finally a protective backing of silicone coating is baked on. Mirrors for optical instruments are usually made by condensing vaporized silver as a film onto glass in a vacuum.

Many industrial brazing and solder alloys are based on silver, e.g. 'easy-flo' which is favoured for brazing copper-based alloys, steels, titanium and many other metals.

Silver bearings, in which silver is plated or bonded onto a steel backing, are used for high-speed and high-load applications such as aero-engines and diesel locomotives.

The antiseptic qualities of silver were mentioned in Chapter 5 and this constitutes the main area of use in medicine. Silver nitrate is sold as solid sticks (lunar caustic) for the treatment of warts and ulcers. Colloidal silver preparations of insoluble forms of silver such as iodides, chlorides or organic proteins ('Protargol') are very effective germicides.

Other uses of the platinum-group metals

It is in their industrial uses that the platinum-group metals show their outstanding qualities.

First and foremost, platinum is an excellent catalyst and it promotes chemical reactions at a rapid rate while itself remaining unchanged. About 30 per cent of the world's production of platinum is used for catalysis in the chemical industry.

Nitric acid is produced by pumping ammonia and air through up to 30 layers of fine platinum—10 per cent rhodium wire mesh to convert them to nitrogen dioxide gas which when dissolved in water gives nitric acid, a key ingredient of modern fertilizers. Platinum catalysts are found in automobile exhausts and in the petroleum-refining industry to convert toxic gases to harmless water vapour and carbon dioxide. They perform a similar function in oxidizing pollutants in a number of industrial plants, e.g. wire enamelling and meat- and fish-processing. One final example of catalytic activity is in aircraft ozone converters on high-flying jet airliners. Ozone has three oxygen atoms in its molecule (O_3). Unlike the life-giving oxygen (O_2), ozone, which exists in the upper reaches of the earth's atmosphere, has toxic properties. The platinum removes one atom from the ozone molecule, giving ordinary oxygen. When the next ozone molecule comes into contact with the converter, it releases that extra atom and this creates two molecules of oxygen.

As we have seen when discussing electroplating, platinum and platinized titanium electrodes are used in electrochemical processes. Examples of industrial use are the electrolysis of brine to make chlorine gas and in fuel cells for producing electricity, e.g. the power source for spacecraft. Other electrical uses are as light-duty contact materials, capacitors and resistors, precision potentiometers, electrodes for heart pacemakers and spark plugs and high-temperature furnace windings.

Platinum and platinum–rhodium alloys play a very important role in temperature measurement. Platinum resistance thermometers rely on the change in electrical resistivity of pure platinum wire with temperature and they have excellent long-term stability over the temperature range -200 to $630°C$. There is a range of platinum/platinum–rhodium thermocouples capable of measuring temperatures up to about $1700°C$.

In glass manufacture, because of their high-temperature capability, oxidation resistance and low risk of contamination, platinum and Pt–Rh or Pt–Ir alloys are used as stirrers, tank blocks and other equipment controlling the flow of glass, as crucibles for melting high-grade optical glass, and as bushings for the extrusion of fibre glass and optical fibres.

Ruby crystals for lasers up to 200 mm in length and 40 mm in width are melted and grown in iridium crucibles.

Platinum-clad anodes are employed for the underwater protection of ships' hulls. By applying a weak electric current, a very slow, almost infinitesimal corrosion of the anodes occurs, preventing attack of the hull.

Razor-blade edges are hardened by sputtering on an ultra-thin layer of a platinum–chromium alloy only a few hundred atoms thick.

The major uses of palladium are as palladium–silver membranes in diffusion units for the production of pure hydrogen and for a range of high-temperature brazing alloys.

The use of osmium and iridium for tipping pen nibs has been mentioned. The fountain pen has largely been replaced by the ball-point pen, but even here we find that the platinum-group metals have adapted to the change. Alloys for the hard ball-points are 0–50 per cent Os, 24–45 per cent Ir, 5–20 per cent Ru, 0–25 per cent Pt, Pd and/or Rh and 0–15 per cent cobalt (Co).

Finally, there is a class of permanent magnets based on Pt–Co alloys; there are the cancer-therapy drugs, cisplatin and neoplatin; and standard weights in Pt–10 per cent Ir alloys are still being produced.

It can be seen from the wide variety of applications given above that the platinum-group metals are of vital importance in today's world.

Bibliography

The following list is a guide to further reading in the areas that have been discussed in this book. Since the book is not directly concerned either with collecting and collections of precious-metal artefacts or with the craft aspects of silversmithing and jewellery, readers are advised to refer to public and institution libraries for publications in these fields.

Bradbury's Book of Hallmarks. J. W. Northend Ltd., 1980.

Butts, A. and Coxe, C. D. (eds.), *Silver: Economics, Metallurgy and Use.* Van Nostrand Reinhold Co. Ltd, London, 1967.

Fischer, J. and Weimar, D. E., *Precious Metal Plating*. Robert Draper Ltd, 1964.

Green, T., *The New World of Gold*. Michael Joseph.

Hughes, G., *The Art of Jewelry*. Studio Vista Publishers, 1972.

Maryon, H., *Metalwork and Enamelling*, 5th edition. Constable and Co. Ltd, London, 1971.

McDonald, D. and Hunt, L. B., *A History of Platinum and its Allied Metals.* Johnson Matthey, 1982.

Rapson, W. S. and Groenewald, T., *Gold Usage*. Academic Press.

Salway, L., *Gold and Gold Hunters*. Kestrel Books, 1978.

Smith, E. A., *Working in Precious Metals*. NAG Press, 1978 (reissue of 1933 edition).

Wise, E. M. (ed.), *Gold: Recovery, Properties and Applications*. Van Nostrand Reinhold Co. Ltd, London, 1964.

Periodicals

Aurum. Published quarterly since 1979 by International Cold Corporation, Geneva, Switzerland.

Gold Bulletin. Published quarterly since 1968 by International Gold Corporation, South Africa.

Platinum Metals Review. Published quarterly since 1957 by Johnson Matthey plc.

Index